PELICAN BOOKS
The Psychology of Learning Mathematics

Richard Skemp was born at Bristol in 1919. He was a foundation scholar at Wellington College, from which he won an open mathematical scholarship to Hertford College, Oxford, in 1937. From 1939 to 1945 he served with the Royal Signals, returning afterwards to Oxford to complete his mathematical degree. During five years of teaching mathematics at secondary schools, he became convinced of the need for more understanding by teachers of how children learn. He went back once more to Oxford and took a psychology degree in 1955.

From then until 1973 he was in the Psychology Department at Manchester University, taking his Ph.D. in 1958, and in charge of the Child Study Unit from 1965. In 1962 he was one of the two British members at the Unesco International Symposium on School Mathematics held in Budapest, and from 1962 to 1969 he directed the Leicestershire Psychology and Mathematics Project. As well as writing school mathematics texts, he has published a number of papers on the psychology of human learning, and is a Fellow of the British Psychological Society. In 1973 he became Professor of Education at Warwick University. In 1978 his title was changed to Professor of Educational Theory, and he became Director of the Mathematics Education Research Centre at Warwick University. From 1980 to 1982 he was President of the International Group for the Psychology of Mathematics Education.

Richard Skemp is married, with one son, and is interested in music, photography, sailing and travel.

D0542189

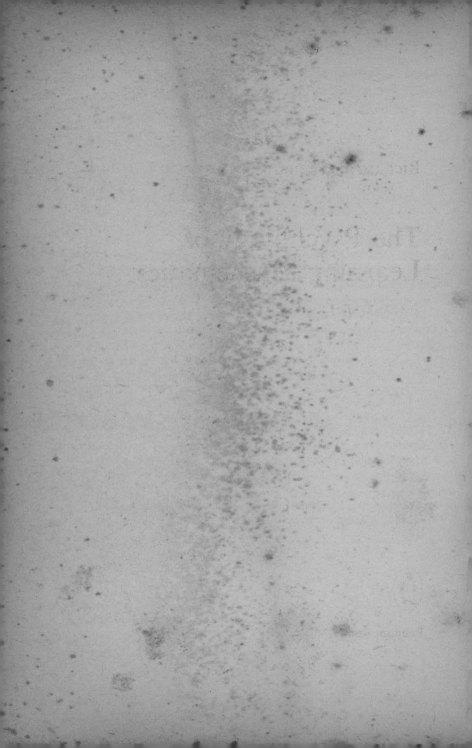

Richard R. Skemp

The Psychology of
Learning Mathematics

SECOND EDITION

Penguin Books

Penguin Books Ltd, Harmondsworth, Middlesex, England
Viking Penguin Inc., 40 West 23rd Street, New York, New York 10010, U.S.A.
Penguin Books Australia Ltd, Ringwood, Victoria, Australia
Penguin Books Canada Ltd, 2801 John Street, Markham, Ontario, Canada L3R 1B4
Penguin Books (N.Z.) Ltd, 182–190 Wairau Road, Auckland 10, New Zealand

First published 1971
Reprinted 1973, 1975, 1977, 1979, 1982
Second edition 1986
Reprinted 1987

Made and printed in Great Britain by
Richard Clay Ltd, Bungay, Suffolk
Set in Monophoto Times

Contents

Acknowledgements 7
Editorial Foreword 9

PART A

Introduction 13
1 The Formation of Mathematical Concepts 19
2 The Idea of a Schema 35
3 Intuitive and Reflective Intelligence 51
4 Symbols 64
5 Different Kinds of Imagery 88
6 Interpersonal and Emotional Factors 108

PART B

Introductory Note 131
7 Beginnings 133
8 The Naming of Numbers 147
9 Two More Key Ideas 162
10 A Need for New Numbers 173
11 Further Expansions of the Number Schema 185
12 Algebra and Problem-solving 213
13 Mappings and Functions 231
14 Generalizing some Geometrical Ideas 250
Retrospect 280

Notes and References 286
Suggestions for Further Reading 288
Index 289

Acknowledgements

Much of the material in Part A of this book was first given at a series of weekend and day courses organized by my friend and colleague Laurie Buxton, at that time Staff Inspector (Mathematics) in the Inner London Education Authority. I am most grateful to him, and to the teachers attending the courses, for this stimulus to my thinking, and for the discussions with teachers which kept me closely in touch with their viewpoints and needs. Laurie Buxton also read the manuscript and made a number of helpful suggestions.

In this, the second, edition, I am glad to have the opportunity to thank the very many persons who have written to me from all over the world – teachers of all kinds, and others who found that, after all, mathematics could and did make sense to them. If I may be forgiven for singling out one from so many, I should like to quote a short extract out of a letter from Beijing which gave me particular pleasure: 'Never didn't I read such good book like this one. It is as though I had found a treasure.'

This and other equally kind letters have been a great enrichment to my life. Thank you all very much.

Editorial Foreword

Mathematics is a curious subject, psychologically. It seems to divide people into two camps, just as there are said to be cat-lovers and dog-lovers; there are those who can do mathematics and there are those who cannot, or who think they cannot, and who 'block' at the first drop of a symbol. Again, mathematicians are said often to be very musical, with the implication that mathematical and musical thinking have something in common. On the other hand, mathematicians are often poor practitioners with words, so supporting the idea that verbal and mathematical intelligence may not go together.

Professor Skemp, the author of this book, is both a mathematician and a psychologist, and, most unusually, he is expert in teaching both subjects. In the first part of the book he looks at the thought processes which people adopt when they do mathematics, and he analyses it psychologically. It is done painlessly so that the reader who knows neither psychology nor mathematics will find himself introduced to a fascinating and important field of inquiry. And those who know something of both subjects will be interested to see the way in which the author has gone beyond previous notions.

In the second part of the book Professor Skemp produces what might almost be thought of as a textbook based on the psychological ideas developed in the first half. It is a bold step, and one which is calculated to give those who already know a little mathematics new insights into their own thinking and into the subject itself. This book will also be a stimulus to all those concerned with improving mathematical teaching.

B. M. FOSS, 1971

PART A

Introduction

Since the early 1960s there has been much concern about the teaching of mathematics, and much activity. In many parts of the world projects have proliferated, and new topics have been introduced under the name of 'modern mathematics' (though most of the topics so described date from before the turn of the century).

Nevertheless, after two decades of effort by many groups of intelligent and hard-working people, so little progress had been made in the U.K. that it was found necessary to set up a governmental committee to inquire into the problems of mathematical education.[1] (Readers elsewhere must judge for themselves whether a similar situation also applies in their own countries.) So where have things gone astray?

Readers for whom mathematics at school was a collection of unintelligible rules which, if memorized and applied correctly, led to 'the right answer' (the criterion for which was a tick by the teacher) would probably agree that there was need for change. Parents of children whose school mathematics also answers this description may feel that there still is. But change is not necessarily for the better, and the introduction of new topics does not bring about better understanding if these are taught in the same bad old way. In contrast, some of us learnt what is now called 'traditional' mathematics in a meaningful way. New topics are not in themselves a sufficient answer.

Other reformers try to present mathematics as a logical development. This approach is laudable in that it aims to show that mathematics is sensible and not arbitrary, but is mistaken in two ways. First, it confuses the logical and the psychological approaches. The main purpose of a logical presentation is to convince doubters; that of a psychological one is to bring about understanding. Second, it gives only the end-product of mathematical discovery ('This is it: all you have to do is learn it'), and fails to bring about in the learner those processes by which mathematical discoveries are made. It teaches mathematical thought, not mathematical thinking.

Problems of learning and teaching are psychological problems, and we can expect little improvement in the teaching of mathematics until we know more about how it is learnt. This book is offered as a contribution mainly in the latter field. It will therefore, I hope, be of interest not only to teachers of mathematics (present and prospective) but also to those who, in spite of past encounters, would still like to learn to understand a little of this subject, and also to parents.

The first part of the book will be concerned with this most basic problem: what *is* understanding, and by what means can we help to bring it about? We certainly think we know whether we understand something or not; and most of us have a fairly deep-rooted belief that it matters. But just what happens when we do understand that does not happen when we don't understand, most of us have no idea. Sometimes, moreover, we think that we have understood something, only to find afterwards that we did not. So until we have a better understanding of understanding itself, we shall be in a poorer position either to understand mathematics ourselves, or to help other people to do so.

In Part B this knowledge will be applied to some fairly basic parts of mathematics; and at the same time the mathematics will be used to illustrate further the ideas developed in Part A. So the first part is mainly about psychology, with occasional references to mathematics; the second part, mainly mathematics, with occasional references to psychology. The mathematics in Part A is there mainly by way of example of psychological ideas, and since other non-mathematical examples are also given, the mathematics here should be taken lightly at any point where it offers difficulty.

Psychology and human learning

Since this is a book about both psychology and mathematics, written by a mathematician turned psychologist, it will be worth discussing the interaction between these two fields of study. For this a few lines of an autobiographical nature are necessary.

I began my professional career as a teacher of mathematics. As my task shifted from that of learning mathematics myself to that of teaching it to other people, I became increasingly concerned with the problem of those pupils who, though intelligent and hard-working, seemed to have a blockage about mathematics. This did not seem to make sense. Surely the main ability required for mathematics was the ability to form and manipulate abstract ideas; and surely this ability coincided closely with what we mean by intelligence? So there seemed to be a contradiction. Gradually I became more and more interested in problems of learning and teaching. These are psychological problems, and to study them more fully, I eventually went back to college and took a degree in psychology. Afterwards, when I was fortunate enough to get a job with opportunities for research, lecturing in psychology at Manchester University, my natural choice for a field of investigation was the problems of learning mathematics.

Habit learning and intelligent learning

A first thread which emerged from these researches was that there seemed to be a qualitative difference between two kinds of learning which we may call habit learning, or rote-memorizing, and learning involving understanding, which is to say intelligent learning. The former can be replicated in the laboratory rat or pigeon, and for various reasons (such as the greater degree of experimental control which is possible) academic psychologists for many years preferred to study this kind of learning. Since the first edition of this book appeared, however, there has been important progress; and influenced by the pioneering work of Piaget, psychologists in many countries are now studying cognitive processes in children and adults.

It is intelligent learning, in contrast to habit learning, in which human beings most excel and in which they most differ from all other species. Though one must not deny intelligence of a sort to the other animals, the intelligence of our own species is so much greater that it has the effect of a difference of kind * and not just of degree. Now, intelligence is a subject which has long been well to the fore in psychological research. But studies in this field have in the main been directed either towards psychometrics (the measurement of intelligence) or towards the relative

* This 'difference of kind' is discussed further in Chapter 1, page 26.

contributions of heredity and environment on intelligence ('nature v. nurture'), while learning theorists still have little to tell us about the interaction between intelligence and learning.

For the psychologist who is interested in intelligent learning, which is to say the formation of conceptual structures communicated and manipulated by means of symbols,* mathematics offers what is perhaps the clearest and most concentrated example. In studying the learning and understanding of mathematics, we are studying the functioning of intelligence in what is at once a particularly pure, and also a widely available, form.

What is intelligence?

What do we mean by intelligence? Though the psychological meaning of this word corresponds roughly to our everyday usage of it, there is as yet no general agreement among psychologists about how it may best be defined.[2] Of those offered by other psychologists, the one which fits in best with my own approach remains that given by Vernon in 1969: 'Intelligence B is the cumulative total of the schemata or mental plans built up through the individual's interaction with his environment, insofar as his constitutional equipment allows.'[3]

Two terms here require further explanation: 'intelligence B' and 'schemata'. The latter will be discussed at length in Chapter 2; briefly, 'schemata' (or 'schemas') means the same as the conceptual structures referred to above. 'Intelligence B' is a term introduced by Hebb in 1949. In his own words: 'From this point of view it appears that the word "intelligence" has *two* valuable meanings. One is (A), an *innate potential*, the capacity for development, a fully innate property that amounts to the possession of a good brain and a good neural metabolism. The second is (B), the functioning of the brain in which development has gone on, determining an *average level of performance or comprehension* by the partly grown or mature person. Neither, of course, is observed directly; but *intelligence B*, a hypothetical level of development in brain function, is a much more direct inference from behaviour than *intelligence A*, the original potential.'[4] Hebb's discussion, from which this passage is extracted, is written from a neurological point of view. The term 'intelligence

---

* The meaning of this sentence will be developed in later chapters.

B' has since become widely used also in the context of mental functioning (as distinct from neural activity).

Mathematics and intelligence B

Mathematics is a particularly good example of intelligence B. There are two reasons for this. First (which is to summarize what has been said already), the learning of mathematics affords many and clear examples of the development of the schemas whose total (including, of course, non-mathematical schemas also) constitutes intelligence B as described by Vernon. Second, the application of mathematics to problems of natural sciences, of technology and of commerce is so powerful that mathematics appears as one of the most, perhaps the most, highly developed mental tools available to us for dealing with our physical environment. If intelligence B is intelligence in its function of understanding, predicting and controlling our physical environment, then mathematics exemplifies intelligence B in one of its most successful developments.

Unrealized potential in Homo sapiens

Our own species, *Homo sapiens*, is a new breakthrough in evolution. But the intelligence wherein lies our superiority remains, in many cultures, largely unrealized. It is no exaggeration to say that the differences in material standards between the technologically most advanced cultures and the most primitive are as great as or greater than the differences between the latter and the highest of the other animal species. The former difference results not from differences in intelligence A, about which the evidence is still quite inconclusive, but from differences in intelligence B, since it is this which is effective in controlling the environment.

But even in those cultures where, through schools, colleges, printing, broadcasting and other means, potential intelligence A is more fully realized as functioning intelligence B, the developmental process is still almost entirely based on tradition and opinion rather than on scientific knowledge and research. If we were able consciously and deliberately to foster the growth of intelligence B, who knows what further strides our civilization might make? And if we want to find out how this might be done, it would be hard to think of a better starting point than to study the learning of mathematics, which increasingly I have come to see as one of the most adaptable and powerful mental tools which human intelligence has constructed, cumulatively over the centuries, for its own use.

This is what the rest of the present book is about. For myself, this starting point has indeed led to greater understanding of human intelligence in a wider context.[5] I hope that it may prove to be so for others also.

CHAPTER 1

The Formation of Mathematical Concepts

Two terms which have recurred throughout the introduction are 'concept' and 'schema'. In this chapter we shall examine what we mean by concepts, and how we form, use and communicate these. Then, in Chapter 2, we shall consider how concepts fit together to form conceptual structures, called schemas, and examine some of the results which follow from the organization of our knowledge into these structures.

Abstracting and classifying

Though the term 'concept' is widely used, it is not easy to define. Nor, for reasons which will appear later, is a direct definition the best way to convey its meaning. So I shall approach it from several directions, and with a variety of examples. Since mathematical concepts are among the most abstract, we shall reach these last.

First, two pre-verbal examples. A baby boy aged twelve months, having finished sucking his bottle, crawled across the floor of the living room to where two empty wine bottles were standing and stood his own empty feeding bottle neatly alongside them. A two-year-old boy, seeing a baby on the floor, reacted to it as he usually did to dogs, patting it on the head and stroking its back. (He had seen plenty of dogs, but had never before seen another baby crawling.)

In both these cases the behaviour of the children concerned implies: one, some kind of classification of their previous experience; two, the fitting of their present experience into one of these classes.

We all behave like this all the time; it is thus that we bring to bear our past experience on the present situation. The activity is so continuous and automatic that it requires some slightly unexpected outcome thereof, such as the above, to call it to our attention.

At a lower level we classify every time we recognize an object as one which we have seen before. On no two occasions are the incoming sense-data likely to be exactly the same, since we see objects at different distances and angles, and also in varying lights. From these varying inputs we abstract certain *in*variant properties, and these properties persist in memory longer than the memory of any particular presentation of the

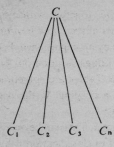

object. In the diagram, C_1, C_2 ... represent successive past experiences of the same object, say, a particular chair. From these we abstract certain common properties, represented in the diagram by C. Once this abstraction is formed, any further experience, C_n, evokes C, and the chair is *recognized*: that is, the new experience is classified with C_1, C_2, etc.; C_n and C are now experienced together; and from their combination we experience both the *similarity* (C) of C_n to our previous experiences of seeing this chair and also the particular distance, angle, etc., on this occasion (C_n).

We progress rapidly to further abstractions. From particular chairs, C, C', C'', we abstract further invariant properties, by which we recognize *Ch* (a new object seen for the first time, say, in a shop window) as a member of this class. It is the second-order abstraction (from the set of abstractions C, C', ...) to which we give the name 'chair'. The invariant

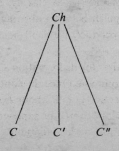

properties which characterize it are already becoming more functional and less perceptual – that is, less attached to the physical properties of a chair. One I saw recently was of basket-work, egg-shaped and hung from a single rope. It bore little or no physical resemblance to any chair which I had ever seen – but I recognized it at once as a chair, and a most desirable one too!

From the abstraction *chair*, together with other abstractions such as *table*, *carpet*, *bureau*, a further abstraction, *furniture*, can be made, and so on. These classifications are by no means fixed. Particularly by the young, chairs are also classified as things to stand on, gymnastic apparatus and framework of a play house. Tables are sometimes used as chairs. This flexibility of classification, according to the needs of the moment, is clearly an aid to adaptability.

Naming an object classifies it. This can be an advantage or a disadvantage. A very important kind of classification is by function, and once an object is thus classified, we know how to behave in relation to it. 'Whatever is this?' 'It's a gadget for pulling off Wellington boots.' But once it is classified in a particular way, we are less open to other classifications. Most of us classify cars as vehicles, time-savers and perhaps status symbols, and use them in accordance with these functions. Fewer also see them as potentially lethal objects, and our behaviour therefore takes less account than it should of this classification.

It may be useful to bring together some of the terms used so far. *Abstracting* is an activity by which we become aware of similarities (in the everyday, not the mathematical, sense) among our experiences. *Classifying* means collecting together our experiences on the basis of these similarities. An *abstraction* is some kind of lasting mental change, the result of abstracting, which enables us to recognize new experiences as having the similarities of an already formed class. Briefly, it is something learnt which enables us to classify; it is the defining property of a class. To distinguish between abstracting as an activity and an abstraction as its end-product, we shall hereafter call the latter a *concept*.

A concept therefore requires for its formation a number of experiences which have something in common. Once the concept is formed, we may (retrospectively and prospectively) talk about *examples* of the concept. Everyday concepts come from everyday experience, and the examples which lead to their formation occur randomly, spaced in time. More frequently encountered objects are, in general, conceptualized more rapidly; but many other factors are at work which make this statement an oversimplification. One of these is *contrast*. In the diagram on the right the single X stands out perceptually from the five variously shaped Os. Objects which thus stand out from their surroundings are more likely to be remembered and their similarities are more likely to be abstracted across intervals of space and time.

The diagram also illustrates the function of non-examples in determining a class. The X, by its difference from all the other shapes, makes the similarity between them more noticeable. The essential characteristics of a *chair* are clarified by pointing to, say, a stool, a settee, a bed and a garden seat, and saying 'These are not chairs.' This is specially useful in fixing the borderline of a class – we use objects which might be examples, but aren't.

Naming

We have just used *naming* again. Language is, in humans, so closely linked with concepts and concept-formation that we cannot for long keep naming out of our discussion. Indeed, many people find it difficult to separate a concept from its name, as is shown by the following charming illustration provided by Vygotsky.[1] Children were told that in a game a dog would be called 'cow'. The following is a typical sequence of questions and answers. 'Does a cow have horns?' 'Yes.' 'But don't you remember that the cow is really a dog? Come now, does a dog have horns?' 'Sure, if it is a cow, if it's called cow, it has horns. That kind of dog has got to have little horns.' Vygotsky also quotes a story about a peasant who, after listening to two students of astronomy talking about the stars, said that he could understand that with the help of instruments people could measure the distance from the earth to the stars and find their positions and motion. What puzzled him was how in the devil they found out the *names* of the stars!

The distinction between a concept and its name is an essential one for our present discussion. A concept is an idea; the name of a concept is a sound, or a mark on paper, associated with it. This association can be formed after the concept has been formed ('What is this called?') or in the process of forming it. If the same name is heard or seen each time, an example of a concept is encountered; by the time a concept is formed, the name has become so closely associated with it that it is not only by children that it is mistaken for the concept itself. In particular, numbers (which are mathematical concepts) and numerals (the names we use for numbers) are widely confused. This point is discussed further in Chapter 4.

Being associated with a concept, the use of a name in connection with an object helps us to classify it, that is, to recognize it as belonging to an existing class. 'What's this?' 'A new kind of bottle opener which works by compressed air.' Now we have classified it, which we were unable to do by its perceptual properties alone; so we know what to do with it. This classification was done by bringing the concept of a bottle opener to consciousness at the same time as the new experience.

Naming can also play a useful, sometimes an essential, part in the formation of new concepts. Hearing the same name in connection with different experiences predisposes us to collect them together in our minds and also increases our chance of abstracting their intrinsic similarities (as distinct from the extrinsic one of being called by the same name). Experiment has also shown that associating different names with classes which are only slightly different in their characteristics helps to classify later examples correctly, even if the later examples are not named. The names help to separate the classes themselves.

The communication of concepts

We can see that language can be used to speed up the formation of a concept by helping to collect and separate contributory examples and non-examples. Can it be used to short-circuit the process altogether by simply defining a concept verbally? Particularly in mathematics, this is often attempted, so let us examine the idea of a definition, as usual with the help of examples. To begin with, let us choose a simple and well-known concept, say, *red*, and imagine that we are asked the meaning of this word by a man, blind from birth, who has been given sight by a corneal graft. The meaning of a word is the concept associated with that word, so our task is now to enable the person to form the concept *red* (which he does not have when we begin) and associate it with the word 'red'.

There are two ways in which we might do this. Being scientifically inclined, and perhaps interested in colour photography, we could give a definition: 'Red is the colour we experience from light of wavelength in the region of 0·6 microns.' Would he now have the concept red? Of course not. Such a definition would be useless *to him*, though not necessarily for other purposes. Intuitively, in such a case, we would point to various objects and say 'This is a red diary, this is a red tie, this is a red jumper . . .' In this way we would arrange for him to have, close together in time, a collection of experiences from which we hope he will abstract the common property – red. Naming is here used as an auxiliary, in the way already described. The same process of abstraction could take place in silence, but it would probably be slower and the name 'red' would not become attached.

If he now asks a different question, 'What does "colour" mean?', we can no longer collect together examples for him by pointing, for the examples we want are *red, blue, green, yellow* . . ., and these are themselves concepts. If, and only if, he already has these concepts in his own mind – their presence in our mind is not enough – then, by collecting together the words for them, we can arrange for him to collect together the concepts themselves, and thus make possible, though not guarantee, the process of abstraction. Naming (or some other symbolization) now becomes an essential factor of the process of abstraction and not just a useful help.

This leads us to an important distinction between two kinds of concept. Those which are derived from our sensory and motor experiences of the outside world, such as *red, motor car, heavy, hot, sweet,* will be called *primary concepts*; those which are abstracted from other concepts will be called *secondary concepts*. If concept A is an example of concept B, then we shall say that B is of a higher order than A. Clearly, if A is an example of B, and B of C, then C is also of higher order than both B and A. 'Of higher order than' means 'abstracted from' (directly or indirectly). So 'more abstract' means 'more removed from experience of the outside world', which fits in with the everyday meaning of the word 'abstract'. This comparison can only be made between concepts in the same hierarchy. Although we might consider that *sonata form* is a more abstract (higher order) concept than *colour*, we cannot properly compare the two.

These related ideas, of order between concepts and a conceptual hierarchy, enable us to see more clearly why, for the person we are thinking of, the definition of red was an inadequate mode of communication: it presupposed concepts such as *colour*, *light*, which could only be formed if concepts such as *red*, *blue*, *green* . . . had already been formed. In general, *concepts of a higher order than those which people already have cannot be communicated to them by a definition* but only by collecting together, for them to experience, suitable examples.

Of what use, then, if any, is a definition?

Two uses can be seen at once. If it were necessary (for example, for a photographic colour filter) to specify exactly within what limits we would still call a colour red, then the above definition would enable us to say where red starts and finishes. And having gone further in the process of abstraction, that is, in the formation of larger classes based on similarities, a definition enables us to retrace our steps. By stating all those (and only those) classes to which our particular concept belongs, we are left with just one possible concept – the one we are defining. In the process we have shown how it relates to the other concepts in its hierarchy. Definitions can thus be seen as a way of adding precision to the boundaries of a concept, once formed, and of stating explicitly its relation to other concepts.

New concepts of a lower order can also be communicated for the first time by this means. For example, if our formerly blind subject asked 'What colour is magenta?' and we could not find a sufficiency of magenta objects to show him, we could say 'It is a colour, between red and blue, rather more blue than red.' Provided that he already had the concepts of blue and red, he could then form at least a beginning of the concept of magenta without ever having seen this colour.

Since most of the new concepts we need in everyday life are of a fairly low order, we usually have available suitable higher-order concepts for the new concepts to be easily communicable by definition, often followed by an example or two, which then serve a different purpose – that of illustration. 'What is a stool?' 'It's a seat without a back for one person' is quite a good definition, but even so a few examples will define the concept in such a way as to exclude hassocks, pouffes and garden swings far more successfully than further elaboration of the definition.

In mathematics, however, not only are the concepts far more abstract than those of everyday life, but the direction of learning is for the most part in the direction of still greater abstraction. The communication of mathematical concepts is therefore much more difficult, on the part of both communicator and receiver. This problem will be taken up again shortly, after certain other general topics have been explored.

Concepts as a cultural heritage

Low-order concepts can be formed, and used, without the use of language.

The criterion for *having* a concept is not being able to say its name but behaving in a way indicative of classifying new data according to the similarities which go to form this concept. Animals behave in ways from which one may reasonably infer that they form simple concepts. A rat, trained to choose a door coloured mid-grey in preference to a light grey, will if now presented with doors of mid-grey and dark grey go to the dark grey. It processes the data in terms of 'darker than'.

The most obvious discontinuity between human beings and other animals is in the former's use of language. What this implies is less obvious. If we choose a word at random it will almost always be found that the concept which the word names – the meaning of the word – is not a specific object or experience but a class. (Proper nouns are a partial exception.)

Now, there are two ways of evoking a concept, that is, of causing it to start functioning. One is by encountering an example of the concept. The concept then comes into action as our way of classifying this example, and our subjective experience is that of *recognition*. The other is by hearing, reading or otherwise making conscious the name, or other symbol, for the concept. Animals can do the first; only human beings can do the second. And the reason for this lies not in superior vocal apparatus, but *in the ability to isolate concepts from any of the examples which give rise to them*. Only by being detachable from the sensory experiences from which they originated can concepts be collected together as examples from which new concepts of greater abstraction can be formed.

We would expect this detachability to be related to abstracting ability, for the stronger the mental organization based not on direct sense-data but on similarities between them, the greater we would expect its ability to function as an independent entity. This view is supported by evidence from several sources. Children of very low intelligence do not learn to talk, in spite of adequate vocal apparatus. Chimpanzees, the closest of our surviving ancestors, can learn to sit at a table and drink from a cup, but not to talk. Human beings are the most intelligent and the most adaptable of all species. They are also the only species who can talk.

Our ability to make concepts independent of the experiences which gave rise to them and to manipulate them by the use of language is the very core of human superiority over other species. This is the first step towards the realization of the potential which this greater intelligence gives. Intelligence makes speech possible, but speech (which has to be learnt) is essential for the formation and use of the higher-order concepts which, collectively, form our scientific and cultural heritage.

A concept is a way of processing data which enables the user to bring past experience usefully to bear on the present situation. Without language each individual has to form their own concepts direct from the environment. Without language, these primary concepts cannot be brought together to form concepts of higher order. By language, however, the first process can be speeded up and the second is made possible. Moreover, the concepts of the past, painstakingly abstracted and slowly accumulated by successive generations, become available to help each new individual form his own conceptual system.

The actual construction of a conceptual system is something which individuals have to do for themselves. But the process can be enormously speeded up if, so to speak, the materials are to hand. It is like the difference between building a boat from a kit of wood already sawn to shape and having to start by walking to the forest, felling the trees, dragging them home, making planks – having first mined some iron ore and smelted it to make an axe and a saw!

What is more, the work of geniuses can be made available to everyone. Concepts like that of gravitation, the result of years of study by one of the greatest intelligences the world has known, become available to all scientists who follow. The first person to form a new concept of this order has to abstract it relatively unaided. Thereafter, language can be used to direct the thoughts of those who follow so that they can make the same discovery in less time and with less intelligence. Yet even Newton (1642–1727) was by no means altogether unaided. He said, with modesty, 'If I have seen a little farther than others, it is because I have stood on the shoulders of giants.' The conceptual structures of earlier mathematicians and scientists were available to him.

In this context, the generalized idea of *noise* is useful. By this is meant data which is irrelevant to a particular communication, so that what is noise in one context may not be so in another. (For example, if we are listening to music when the telephone rings, the sound of the bell conveys *information* that someone is calling us, but is *noise* relative to the music.) The greater the noise, the harder it is to form the concept. Before reading on, please put your hand over the diagrams which are on the right-hand side at the foot of this page. Try to form the concept from the high-noise examples and non-examples. Now remove your hand and try to form the concept from the low-noise examples of the same concept.

From the right-hand examples it is much easier to see that the concept is *having intersecting lines*. The extra noise in the left-hand examples comes partly from the additional lines, but largely from the fact that each looks like something.

An attribute of high intelligence is the ability to form concepts under conditions of great noise. But once we have a concept, we can see examples of it where previously we could not.

High noise   Low noise

Examples   Examples

Not examples   Not examples

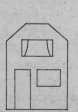

The power of conceptual thinking

Conceptual thinking confers on users great power to adapt their behaviour to the environment, and to shape their environment to suit their own requirements. This results partly from the detachment of the concepts from both present sense-data and behaviour, and their manipulation independently of these. We take this so much for granted that we hardly realize the enormous advantage of *not* having to do something in order to discover whether it is the best thing to do! But, of course, all major activities, from setting up in business to building an aircraft, are put together in thought before they are constructed in fact.

The power of concepts also comes from their ability to combine and relate many different experiences and classes of experience. The more abstract the concepts, the greater their power to do this. The person who says 'Don't worry me with theory – just give me the facts' is speaking foolishly. A set of facts can be used only in the circumstances to which they belong, whereas an appropriate theory enables us to explain, predict and control a great number of particular events in the classes to which it relates.

A further contribution to the power of conceptual thinking is related to the shortness of our span of attention. Our short-term memory can only store, on average, seven words or other symbols. (The range usually quoted is 7 ± 3.) Clearly the higher the order of the concepts which these symbols represent, the greater the stored experience they bring to bear. Mathematics is the most abstract, and so the most powerful, of all theoretical systems. It is therefore potentially the most useful; scientists in particular, but also economists and navigators, businessmen and communications engineers, find it an indispensable 'tool' (data-processing system) for their work.

Its usefulness is, however, only potential, and many who work wearily at trying to learn it throughout their schooldays derive little benefit, and no enjoyment. This is almost certainly because they are not really learning mathematics at all. The latter is an interesting and enjoyable process, though many will find this hard to believe. What is inflicted on all too many children and older students is the manipulation of symbols with little or no meaning attached, according to a number of rote-memorized rules. This is not only boring (because meaningless); it is very much harder, because unconnected rules are much harder to remember than an integrated conceptual structure. The latter point will be taken up in the next chapter. Here we shall concentrate on the communication of mathematical concepts.

The learning of mathematical concepts

Much of our everyday knowledge is learnt directly from our environment, and the concepts involved are not very abstract. The particular problem (but also the power) of mathematics lies in its great abstractness and generality, achieved by successive generations of particularly intelligent individuals each of whom has been abstracting from, or generalizing, concepts of earlier generations. The present-day learner has to process not raw data but the data-processing systems of existing mathematics. This is not only an immeasurable advantage, in that an able student can acquire in years ideas which took centuries of past effort to develop; it also exposes the learner to a particular hazard. Mathematics cannot be learnt directly from the everyday environment, but only indirectly from other mathematicians, in conjunction with one's own reflective intelligence. At best, this makes one largely dependent on teachers (including all who write mathematical textbooks); at worst, it exposes one to the possibility of acquiring a lifelong fear and dislike of mathematics.

Though the first principles of the learning of mathematics are straightforward, it is the communicator of mathematical ideas, and not the recipient, who most needs to know them. And though they are simple enough in themselves, their mathematical applications involve much hard thinking. The first of these principles was stated earlier in the chapter:

(1) *Concepts of a higher order than those which people already have cannot be communicated to them by a definition, but only by arranging for them to encounter a suitable collection of examples.*

The second follows directly from it:

(2) *Since in mathematics these examples are almost invariably other concepts, it must first be ensured that these are already formed in the mind of the learner.*

The first of these principles is broken by the vast majority of textbooks, past and present. Nearly everywhere we see new topics introduced not by examples but by definitions, definitions of the most admirable brevity and exactitude for the teacher (who already has the concepts to which they refer) but unintelligible to the student. For reasons which will be apparent, examples cannot be quoted here, but readers are invited to verify this statement for themselves. It is also a useful exercise to look at some definitions of ideas new to oneself in books about mathematics beyond the stage which one has reached. This enables one to experience at first hand the bafflement of the younger learner.

Good teachers intuitively help out a definition with examples. To choose a suitable collection is, however, harder than it sounds. The examples must have in common the properties which form the concept but no others. To put it differently, they must be alike in the ways which are to be abstracted, and otherwise different enough for the properties irrelevant to this particular concept to cancel out or, more accurately, fail to summate. Remembering that these irrelevant properties may be regarded as noise, we may say that some noise is necessary to concept formation. In the earlier stages, low noise – clear embodiment of the concept, with little distracting detail – is desirable; but as the concept becomes more strongly established, increasing noise teaches the recipient to abstract the conceptual properties from more difficult examples and so reduces dependence on the teacher.

Composing a suitable collection thus requires both inventiveness and a very clear awareness of the concept to be communicated. Now, it is possible to have, and use, a concept at an intuitive level without being consciously aware of it.* This applies particularly to some of the most basic and frequently used ideas: partly because the more automatic any activity, the less we think about it; partly because the most fundamental ideas of mathematics are acquired at an early age, when we have not the ability to analyse them; and partly because some of these fundamental ideas are also among the most subtle. But it is easy to slip up even when these factors do not apply.

Some children were learning the theorem of Pythagoras (c. sixth century BC). They had copied a right-angled triangle from the blackboard – figure a – and were told to make a square on each side. This they did easily enough for the two shorter sides – figure b; but they were nearly all in difficulty when they tried to draw the square on the hypotenuse. Many

* This is expanded in Chapter 3.

of them drew something like figure c. From this, I inferred that the squares from which they had formed their concepts had all been 'square' to the paper and had included no obliquely placed examples. All too easily done!

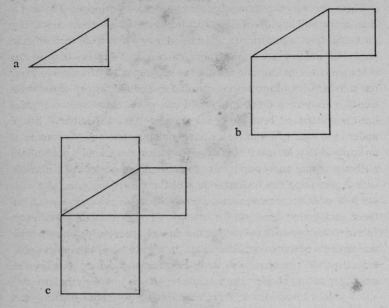

The second of the two principles, that the necessary lower-order concepts must be present before the next stage of abstraction is possible, seems even more straightforward. To put this into effect, however, means that before we try to communicate a new concept, we have to find out what are its contributory concepts; and for each of these, we have to find out *its* contributory concepts, and so on, until we reach either primary concepts or experience which we can assume. When this has been done, a suitable plan can then be made which will present to the learner a possible, and not an impossible, task.

This conceptual analysis involves much more work than just giving a definition. If done, it leads to some surprising results. Ideas which not long ago were first taught in university courses are now seen to be so fundamental that they are being introduced in the primary school: for example, sets, one-to-one correspondence.* Other topics still regarded as

* See Chapter 7, pages 133 and 138.

elementary are found on analysis to involve ideas which even those teaching the topic have for the most part never heard of. In this category I include the manipulation of fractional numbers.

There are two other consequences of the second principle. The first is that in the building up of the structure of successive abstractions, if a particular level is imperfectly understood, everything from then on is in peril. This dependency is probably greater in mathematics than in any other subject. One can understand the geography of Africa even if one has missed that of Europe; one can understand the history of the nineteenth century even if one has missed that of the eighteenth; in physics one can understand 'heat and light' even if one has missed 'sound'. But to understand algebra without ever having really understood arithmetic is an impossibility, for much of the algebra we learn at school is generalized arithmetic. Since many pupils learn to do the manipulations of arithmetic with a very imperfect understanding of the underlying principles, it is small wonder that mathematics remain a closed book to them. Even those who get off to a good start may, through absence, inattention, failure to keep up with the pace of the class or other reasons, fail to form the concepts of some particular stage. In that case, all subsequent concepts dependent on these may never be understood, and pupils become steadily more out of their depth. In the latter case, however, the situation may not be so irremediable, if the learning situation is one which makes back-tracking possible: for example, if the text in use provides a genuine explanation and is not just a collection of exercises. Success will then depend partly on the confidence of the learners in their own powers of comprehension.

The other consequence (of the second principle) is that the contributory concepts needed for each new stage of abstraction must be *available*. It is not sufficient for them to have been learnt at some time in the past; they must be accessible when needed. This is partly a matter, again, of having facilities available for back-tracking. Appropriate revision, planned by a teacher, will be specially useful for beginners, but more advanced students should be taking a more active part in the direction of their own studies, and, for these, returning to take another look at earlier work will be more effective if it is directed by a felt need rather than by an outside instruction. To put it differently, an answer has more meaning to someone who has first asked a question.

Learning and teaching

In learning mathematics, although we have to create all the concepts anew in our own minds, we are only able to do this by using the concepts arrived at by past mathematicians. There is too much for even a genius to do in a lifetime.

This makes the learning of mathematics, especially in its early stages and for the average student, very dependent on good teaching. Now, to know mathematics is one thing and to be able to teach it – to communicate it to those at a lower conceptual level – is quite another; and I believe that it is the latter which is most lacking at the moment. As a result, many people acquire at school a lifelong dislike, even fear, of mathematics.

It is good that widespread efforts have been and are still being made to remedy this, for example, by the introduction of new syllabi, more attractive presentation, television series and other means. But the small success of these efforts, after twenty years or more, supports the view already put forward in the introduction, namely; that these efforts will be of little value until they are combined with greater awareness of the mental processes involved in the learning of mathematics.

*

In this chapter, though we have been discussing the formation of mathematical concepts, most of the examples used have been non-mathematical. The concepts of mathematics are the result of so many abstractions, derived from abstractions, derived from abstractions, and so on, that the psychological argument would have been in danger of being lost in the complexity of the mathematical examples. Even such topics as counting or long multiplication are found on examination to involve surprisingly many lower-order concepts.

Readers who feel that it is time for some mathematical examples may, if they wish, now turn to Chapter 7, the first chapter of Part B. There, the ideas of this chapter are applied in an examination of some of the most basic ideas of elementary mathematics. It is suggested that Chapter 2 be read then, before continuing further with Part B, after which readers may read parts A and B in parallel or read the book consecutively as printed, according to their own preference.

The Idea of a Schema

Though in the previous chapter our attention was centred on the formation of single concepts, each of these by its very nature is embedded in a structure of other concepts. Each (except primary concepts) is derived from other concepts and contributes to the formation of yet others, so it is part of a hierarchy. But at each level alternative classifications are possible, leading to different hierarchies. A car can be classed as a vehicle (with buses, trains, aircraft), as a status symbol (with a title, a good address, a mink coat), as a source of inland revenue (with tobacco, drink, and dog licences), as an export (with gramophone records, Scotch whisky, Harris tweed), etc. What is more, the class concepts on which we have been concentrating so far are by no means the only kind. Given a collection not of single objects but of *pairs* of objects we may become aware of something in common between the pairs. For example:

puppy, dog; kitten, cat; chicken, hen.

Here we see that each of these pairs can be connected by the idea '... is a young ...' Another example:

Bristol, England; Hull, England; Rotterdam, Holland.

In this, each pair can be connected by the idea '... is a port of ...' These two connecting ideas are themselves examples of a new idea called a *relation*.

A mathematical relation may be seen in the following collection of pairs.

(6, 5), (2, 1), (9, 8), (32, 31) ...

We can call this relation 'is one more than' or 'is the successor of'. Another mathematical example:

$$(\tfrac{1}{2}, \tfrac{2}{4}), (\tfrac{1}{3}, \tfrac{2}{6}), (\tfrac{1}{4}, \tfrac{2}{8}) \ldots$$

This relation is called 'is equivalent to'. The fractions in each pair, though not identical, represent the same number.* Notice (1) that in mathematics it is usual to enclose the pairs in a given relation in parentheses, as above; (2) that the order within the pairs usually matters. These:

$$(5, 6), (1, 2), (8, 9), (31, 32)$$

are in a different relation to these:

$$(6, 5), (2, 1), (9, 8), (32, 31)$$

We can even start to classify these relations. Those mathematical relations given as examples in the last paragraph were chosen to exemplify two particular kinds: order relations and equivalence relations. Other order relations are: is greater than, is the ancestor of, happened after. Other equivalence relations are: is the same size as, is the sibling of, is the same colour as. Both order relations and equivalence relations have important general properties. So we have not only a hierarchical structure of class concepts but another structure of individual relations, and classes of relations, which forms cross-linkages within the first structure.

Another source of cross-linkages arises from our ability to 'turn one idea into another' by doing something to it.

Example: good → bad   hot → cold   high → low

Another example: good → best   bad → worst   high → highest

This 'something which we can do to an idea' is called a *transformation*, or more generally a *function*. There are many different kinds of transformation, and, what is more, we can on occasion combine two particular transformations to get another transformation (just as one can combine two numbers to get another). For example, by combining the two transformations above we get

good → worst,   hot → coldest,   etc.

So transformations are both connected among themselves and are also another source of connections between the ideas to which they can be applied.

The foregoing offers a brief, and perhaps rather concentrated, glimpse of the richness and variety of the ways in which concepts can be inter-

---

* This is explained further in Chapter 10.

related, and of the resulting structures. The study of the structures them-selves is an important part of mathematics, and the study of the ways in which they are built up and function is at the very core of the psychology of learning mathematics.

Now, when a number of suitable components are suitably connected, the resulting combination may have properties which it would have been difficult to predict from a knowledge of the properties of the individual components. How many of us could have predicted from knowledge of the separate properties of transistors, condensers, resistors and the like that, when these are suitably connected, the result would enable us to hear radio broadcasts?

So it is with concepts and conceptual structures. The new function of the electrical structure described above is marked by a new name – transistor radio. Likewise, a conceptual structure has its own name – *schema*. The term includes not only the complex conceptual structures of mathematics but also relatively simple structures which coordinate sensori-motor activity. Here we shall be concerned mainly with abstract conceptual schemas. The previous chapter has shown that these concepts have their origins in sensory experience of, and motor activity towards, the outside world. But they soon become detachable from their origins, and their further development takes place by interaction with other mathematicians and with each other.

Among the new functions which a schema has, beyond the separate properties of its individual concepts, are the following: it integrates exist-ing knowledge, it acts as a tool for future learning and it makes possible understanding.[1]

The integrative function of a schema

When we recognize something as an example of a concept we become aware of it at two levels: as itself and as a member of this class. Thus, when we see some particular car, we automatically recognize it as a member of the class of private cars. But this class-concept is linked by our mental schemas with a vast number of other concepts, which are available to help us behave adaptively with respect to the many different situations in which a car can form a part. Suppose the car is for sale. Then all our motoring experience is brought to bear, reviews of its performance may be recalled, questions to be asked (m.p.g.?) present themselves. Suppose that the cost is beyond our present bank balance. Then sources of finance (bank loans, hire purchase) come to mind. Suppose, instead, that the car is on the road, but has broken down. Then instruments of help (such as the A.A., nearest garage, telephone boxes) are recalled.

Most of these schemas have probably already been linked with the car concept in the past. But suppose now that we park on a foreshore and find that our wheels have sunk into the soft sand. This presents a problem, to solve which schemas from other fields of experience must be brought to bear, such as the behaviour of tides, ways of making a firm surface on soft sand. The more other schemas we have available, the better our chance of coping with the unexpected. We shall return to this point later in the chapter.

The schema as a tool for further learning

Our existing schemas are also indispensable tools for the acquisition of further knowledge. Almost everything we learn depends on knowing something else already. To learn aircraft designing we must know aerodynamics, which depends on prior knowledge of calculus, which requires knowledge of algebra, which depends on arithmetic. To learn advanced physiology requires biochemistry, which needs a knowledge of elementary 'school' chemistry. These, and all higher learning, depend on the basic schemas of reading, writing and speaking (or, exceptionally, communicating in some other way) our native language.

This principle – the dependency of new learning on the availability of a suitable schema – is a generalization of the second principle for conceptual learning, stated in Chapter 1 on page 30. In the generalized form, new features become important which were not so noticeable while we were concentrating on the learning of particular concepts, though using hindsight they can be seen to be latent there. As an introduction to these, it will be useful to look at an experiment [2] whose purpose was to try to isolate the factor of a schema in learning, or more precisely, to find out how much difference the presence or absence of a suitable schema made to the amount of new material which could be learnt.

For the purpose of the experiment, an artificial schema was devised, somewhat resembling a Red Indian sign language. On the first day the subjects learnt the meanings of sixteen basic signs, such as:

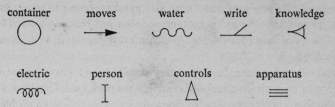

On the second day meanings were assigned to pairs or trios of symbols, such as:

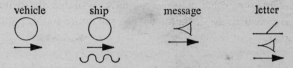

The meanings of these small groups of symbols are related to the meanings of the single symbols, as the reader can verify. On the third and fourth days the groups to be learnt were made progressively larger, the meanings again being related to those of the smaller groups. Here are some examples. (Note that (( )) means plural.)

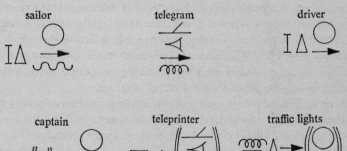

The final task, on the fourth day, was to learn two pages of symbols, each page containing a hundred symbols in ten groups each having from eight to twelve symbols. On one page each group was given a meaning related to the meanings of the smaller groups, as in the examples given. The other page contained groups which were in fact similarly meaningful to a comparison group, but not to these subjects. The comparison group had learnt the same symbols but with different meanings, and these had been built up into a different schema. So in their final task each group had an appropriate schema for one page and an inappropriate schema for the other page. In other words, what was meaningful (in terms of earlier learning) to one group was non-meaningful to the other, and vice versa.

When the results of schematic and 'rote' learning were compared, the differences were striking.

	% recalled (all subjects)		
	Immediate	*After one day*	*After four weeks*
Schematic	69	69	58
Rote	32	23	8

In this case twice as much was recalled of the schematically learnt as of the rote-learnt material when tested immediately afterwards; and in four weeks the proportion had changed to seven times as much. The schematically learnt material was not only better learnt, but better retained.

Objectively, the two pages of symbols were the same for all the subjects. The only difference was in the mental structures which they had available for the learning task. Clearly, therefore, the schemas which we build up in the course of our early learning of a subject will be crucial to the ease or difficulty with which we can master later topics. When learning schematically – which, in the present context, is to say intelligently – we are not only learning much more efficiently what we are currently engaged in; we are preparing a mental tool for applying the same approach to future learning tasks in that field. Moreover, when subsequently using this tool, we are consolidating the earlier content of the schema. This gives schematic learning a triple advantage over rote memorizing.

There are, however, also certain possible disadvantages to be considered.

The first is that, if a task is considered in isolation, schematic learning may take longer. For example, rules for solving a simple equation (see page 49) can be memorized in much less time than it takes to achieve understanding. So if all one wants to learn is how to do a particular job, memorizing a set of rules may be the quickest way. If, however, one wishes to progress, then the number of rules to be learnt becomes steadily more burdensome until eventually the task becomes excessive. A schema, even more than a concept, greatly reduces cognitive strain. Moreover, in most mathematical schemas, all the main contributory ideas are of very general application in mathematics. Time spent in acquiring them is not only of psychological value (meaning that present and future learning is easier and more lasting) but of mathematical value (meaning that the ideas are also of great importance mathematically). In the present context, good psychology is good mathematics.

The second disadvantage is more far-reaching. Since new experience which fits into an existing schema is so much better remembered, a schema has a highly selective effect on our experience. What does *not* fit into it is largely not learnt at all, and what is learnt temporarily is soon forgotten. So, not only are unsuitable schemas a major handicap to our future learning, but even schemas which have been of real value may cease to become so if new experience is encountered, new ideas need to be acquired, which cannot be fitted in to an existing schema. A schema can be as powerful a hindrance as help if it happens to be an unsuitable one.

This brings us to a consideration of adaptability at a new level. So far a schema has been seen as a major instrument of adaptability, being the most effective organization of existing knowledge both for solving new problems and for acquiring new knowledge (and thereby for solving still more new problems in the future). But its very strength now appears as its potential downfall, in that a strong tendency emerges towards the self-perpetuation of existing schemas. If situations are then encountered for which they are not adequate, this stability of the schemas becomes an obstacle to adaptability. What is then necessary is a change of structure in the schemas: they themselves must adapt. Instead of a stable, growing schema by means of which the individual organizes past experience and *assimilates* new data, *reconstruction* is required before the new situation can be understood. This may be difficult, and if it fails, the new experience can no longer be successfully interpreted and adaptive behaviour breaks down – the individual cannot cope.

An everyday example will illustrate the idea, after which some mathematical examples will be given. Early in life, a boy learns to distinguish between his compatriots and foreigners. His schema of a foreigner is that of a person who comes from abroad, who speaks English with a different accent from his own, perhaps only with difficulty, whose own language is novel and usually incomprehensible, whose mode of dress and personal appearance are slightly or very different. New individual foreigners and new classes – people from countries he had never heard of – are easily assimilated to this concept, which leads to expansion of his schema. But suppose now that he takes a holiday abroad with his parents and discovers that he himself is described as a foreigner. To him, this is incomprehensible. The local inhabitants are the foreigners; he is British! Before he can comprehend this new experience – assimilate it to his schema – the schema itself has to be restructured. His idea of foreigners has to become that of people in a country which is not their own. Not only does this new concept enable him to understand the new experience and so to behave appropriately; it includes the earlier concept as a special case. This is the best kind of reconstruction.

A schema is of such value to an individual that the resistance to changing it can be great, and circumstances or individuals imposing pressure to change may be experienced as threats – and responded to accordingly. Even if it is less than a threat, reconstruction can be difficult, whereas assimilation of a new experience to an existing schema gives a feeling of mastery and is usually enjoyed.

One of the most basic mathematical schemas which we learn is that of the natural number system – the set of counting numbers together with the operations of addition and multiplication. Having learnt to count to ten, a child rapidly progresses to twenty, and is eager to continue the process. Adding single-figure numbers, with the help of concrete materials, is soon learnt. Extending this to the addition of two-figure numbers requires, first, an understanding of our system of numeration based on place value, but once this has been mastered, addition of three-, four-, five-figure numbers is again a straightforward extension. Multiplication is like repeated addition, long multiplication extends simple multiplication. Throughout, the process is one of expansion.

It is another matter when fractional numbers are encountered. These constitute a new number system, not an extension of one which is known already. The system of numeration is different in itself and has new characteristics: for example, an infinite number of different fractions can be used to represent the same number. Multiplication can no longer be understood in terms of repeated addition. Before fractional numbers can be understood, a major reconstruction of the number schema is required. Some people, indeed, go through life without ever really understanding fractional numbers, and small blame to them. Their teacher probably never understood them either, and the difficulty of this particular reconstruction is such that it would require a child of genius level to achieve it unaided at the age when this task is encountered.

The history of mathematics contains some interesting examples showing the difficulty of reconstruction presented by a new number system. When Pythagoras discovered that the length of the hypotenuse of a right-angled triangle could not always be expressed as a rational number, he swore the members of his school to secrecy about this threat to their existing ways of thinking. In his well-known history of mathematics,[3] E. T. Bell says: 'When negative numbers first appeared in experience, as in debits instead of credits, they, as *numbers*, were held in the same abhorrence as "unnatural" monstrosities as were later the "imaginary" numbers $\sqrt{-1}, \sqrt{-2}$, etc.' The Hindu-Arabic system of numerals for the natural numbers also met with great resistance when it was first introduced into Europe in the thirteenth century, and in some places its use was even made illegal. Unspeakable, unnatural, illegal – these are the ways in which the ordinary working tools of present-day mathematics were all characterized by some of the mathematicians who first encountered them. But now that we know the importance of our personal schemas to us, we can begin to understand the defensive nature of these reactions to any new ideas which threaten to overthrow them.

Understanding

We are now in a position to say what we mean by understanding. *To understand something means to assimilate it into an appropriate schema.* This explains the subjective nature of understanding and also makes clear that this is not usually an all-or-nothing state. We may achieve a subjective feeling of understanding by assimilation to an inappropriate schema – the Greeks 'understood' thunderstorms by assimilating these noisy affairs to the schema of a large and powerful being, Zeus, getting angry and throwing things. In this case, an appropriate schema involves the idea of an electric spark, so it was not until the eighteenth century that any real understanding of thunderstorms was possible. The first and major step was taken by Benjamin Franklin, who assimilated concepts

about thunderstorms to those about electrical discharges. Fuller understanding, however, involves knowledge of ionization processes in the
atmosphere – assimilation to a more extensive schema. What happens in
a case like this is that the basic schema becomes enlarged and to the
original points of assimilation – noise to noise, lightning flash to electric
spark – more are added. Better internal organization of a schema may
also improve understanding, and clearly there is no stage at which this
process is complete. One obstacle to the further increase of understanding
is the belief that one already understands fully.

We can also see that the deep-rooted conviction mentioned earlier,
that it matters whether or not we understand something, is well-founded.
For this subjective feeling that we understand something, open to error
though it may be, is in general a sign that we are therefore now able to
behave appropriately in a new class of situations.

The difference in adaptability between that based on a rule and that
which results from understanding has been well demonstrated experimentally by M. A. Bell.[4] The example was chosen from topology, a
branch of mathematics which will perhaps be new to readers, who may, if
they wish, try it for themselves. It has the advantage that the relevant
schema can be quickly built up, whereas most mathematical ones take
longer.

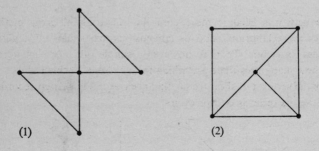

(1)   (2)

These two diagrams represent topological networks, which are made
up of points called *vertices* joined by straight or curved lines called *arcs*.
To *traverse* a network means to follow a continuous path covering every
arc of the network once and only once. A few trials will show that
network (1) can be traversed, whereas (2) cannot. Here are two more
examples.

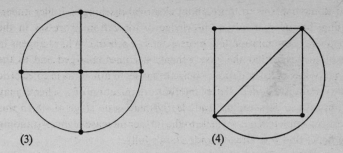

(3)   (4)

By trial and error, it is easy to find that network (4) can be traversed, and the reader will soon become convinced that (3) cannot, though this is not the same as proving that it is impossible.

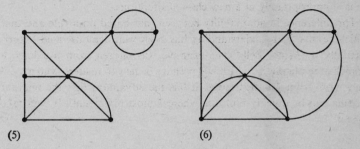

(5)   (6)

As the networks become more complex, the trial-and-error method becomes more laborious, and its conclusions, particularly if negative, carry less conviction. There is, however, a simple rule. For each vertex, count how many arcs there are which meet there. Call this number the *order* of the vertex. For short we say that a vertex is odd or even according to whether its order is odd or even.

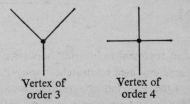

Vertex of   Vertex of
order 3   order 4

Rule: a network can be traversed if, and only if, the number of odd vertices is zero or two.

With this rule it is easy to verify that network (6) can be traversed, starting in the top left-hand corner, and that (5) cannot. More complicated networks present little greater difficulty.

Two groups of eleven-year-old children were introduced to the above ideas. Group 1 was given the rule and also an explanation (which will be withheld from the reader at this stage) of the reason for the rule. Group 2 was given just the rule. Both groups of children were then given twelve problems of this kind, including some quite complicated networks. All children of both groups got all the problems right. At this stage, one could not distinguish by their results between children who understood the reason for the rule and those who did not.

A further set of network problems was then presented to these two groups, with one small difference. Here are four typical networks from the set.

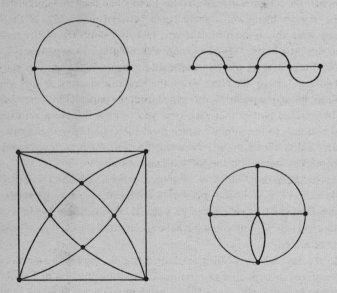

The new problem was (a) to try to find which networks could be traversed as before, but ending this time at the starting point; and (b) to try to find a rule for doing this. Before reading further, readers may care to try for themselves.

A third group of children, with no previous experience of these problems and no knowledge of the rule, was also given this new task. The results, in terms of children finding the correct new rule, were:

Group 1 (first rule with understanding)	nine children out of twelve	(75%)
Group 2 (first rule without understanding)	three children out of ten	(30%)
Group 3 (no previous knowledge)	two children out of twelve	(17%)

Whereas the earlier results of Groups 1 and 2 had been indistinguishable, these new problems show a great gap between them. 75% of the first group were able to adapt to the new task, but only 30% of the second, who did little better than the group with no previous experience.

Now take a sheet of plain paper and copy on it the vertices only of network (1), page 44. Next, draw the network starting at any vertex without lifting the point of your pencil from the paper. (This corresponds to traversing.) Notice that each time you enter and leave a vertex, you add two arcs to the number which meet there, that is, you increase the order of that vertex by two. Now do the same for network (4) and for network (6), starting in the top left-hand corner.

This explanation, which is, of course, briefer than that given to the children, will, it is hoped, provide a sufficient clue for the reader to understand the first rule, given on page 45. If you succeeded in finding the second rule without this explanation, congratulations! If not, it should now be easier.

In the days when so-called teaching-machines were being marketed, I came across an expensive programme called 'Introduction to Topology', published for use with an expensive teaching-machine, in which the first rule (only) was given, and without explanation. In this form, it is not only hard to adapt to problems of the second kind; it does not enable one to answer other relevant questions like 'How can we be sure that this rule works for *any* network?', 'Would it work for a three-dimensional network?' and especially 'How can we be sure that a given network *cannot* be traversed by somebody clever enough?' All these questions can be answered by someone who has understood the explanation of the rule, thereby demonstrating further the greater adaptability of the schema to new problems.

Implications for the learning of mathematics

The central importance of the schema as a tool of learning means that inappropriate early schemas will make the assimilation of later ideas much more difficult, perhaps impossible.

'Inappropriate' also includes non-existent. Learning to manipulate symbols in such a way as to obtain the approved answer may be very hard to distinguish in its early stages from conceptual learning. The learner cannot distinguish between the two if he has no experience of genuinely understanding mathematics. And all the teachers can see (or hear) are the symbols. Not being thought-readers, they have no direct knowledge of whether or not the right concepts, or any at all, are attached. The way to find out is to test the adaptability of the learner to new, though mathematically related, situations. Mechanical computation does not do this. The amount which a bright child can memorize is remarkable, and the appearance of learning mathematics may be maintained until a level is reached at which only true conceptual learning is adequate to the situation. At this stage the learner tries to master the new tasks by the only means known – memorizing the rule for each kind of problem. This task being now impossible, even the outward appearance of progress ceases, and, with accompanying distress, another pupil falls by the way-side.

An appropriate schema means one which takes into account the long-term learning task and not just the immediate one. The solution of equations, for example, is sometimes based on the idea of a pair of scales. If we add or subtract the same weights in each pan, the balance is preserved, and thus we can find a weight which exactly balances the unknown weight. This model also justifies 'taking a number to the other side and changing the sign', since we get the same result from adding, say, 3 kg. to the left-hand pan, or taking it away from the right-hand pan.

At the beginning stages, this simple schema is admirable. It does have the disadvantage of regarding x as an unknown quantity which we have to 'find', which is not a basic concept of mathematics, instead of as a variable, which is. But its chief defect is that the schema of 'balancing the two sides' does not apply to equations like

$$x + 4 = 0$$
$$x^2 = 4$$
and
$$x^2 - 3x = 4$$

except by stretching it till it creaks, and by no device to

$$x^2 + 4 = 0$$
and
$$\frac{dy}{dx} = 4$$

The teacher must look far beyond the present task of the learner, and wherever possible communicate new ideas in such a way that appropriate long-term schemas are formed.

In spite of its shortcomings, the above schema is still incomparably better than the collection of rules without reasons which are sometimes taught, since it does make sense and therefore contributes to an overall belief in mathematics as a meaningful activity. It may also be difficult sometimes to choose between an easy but short-term initial schema and a harder long-term one. This is not quite the same kind of choice as that which we may face when shopping, between something cheap but short-lasting and something more expensive but longer-lasting, since we cannot throw away our early schemas. We have to reconstruct them, which, as we have seen, may present difficulties. So the choice may not always be an easy one. In general, however, it often happens that the more general, long-term ideas are not necessarily harder to learn but only harder to find initially. This transfers the difficulty from the learner to the teacher.

The responsibility of teachers in the early stages of learning is therefore great. They have to make sure that schematic learning, not just memorizing the manipulations of symbols, is taking place. They have to know which stages require only straightforward assimilation and when reconstruction is needed, since at the latter stages the pace must be slower and progress more carefully checked. And they have to plan on a long-term basis the schemas which will be most adaptable to future as well as present needs.

To satisfy fully the latter requirement is impossible. The present rate of change in mathematics, and the uses to which it is put, is such that no one can know what future tasks the present learners of mathematics will have to face. And the rate of change is increasing. So what is best to do?

The first part of the answer would seem to be to try to lay a well-structured foundation of basic mathematical ideas on which the learner can build in whatever future direction becomes necessary: that is, to find for oneself, and help one's pupils to find, the basic patterns; secondly, to teach them always to be looking for these for themselves; and thirdly, to teach them always to be ready to reconstruct their schemas – to appreciate the value of these as working tools, but always to be willing to replace them by better ones. The first of these is teaching mathematics; the second and third are teaching pupils to learn mathematics. Only these last two prepare pupils for an unknown future.

CHAPTER 3

Intuitive and Reflective Intelligence

There is an anecdote about a very well-known professor of mathematics which, if it is not true, deserves to be. It relates that, while addressing a learned audience, he wrote a mathematical statement on the board, saying 'This, of course, is obvious.' Looking at it again, he said 'At least, I think it is obvious.' Growing more doubtful, he said 'Excuse me,' and taking pencil and paper, was absent from the room for about twenty minutes. He returned beaming and said triumphantly 'Yes, gentlemen, it *is* obvious.'

Psychologically, the charm of this story is that there is no inconsistency between the first confident statement and the relatively lengthy period of deliberation needed, once doubt had arisen, before this confidence could be regained. By the first statement the speaker meant 'We can accept intuitively the truth of this statement.' By the second statement he meant that, having subjected it to logical analysis, he had confirmed that this intuitive acceptance was justified. Being sure of something is one thing; knowing why one is sure is another.

A similar example. Multiply 16 by 25. (i) What is the answer? (ii) Now explain how you did it. To answer the second question involves turning your attention from the task itself to your mental processes involved in doing it.

Another example. 'What I am writing with is chalk.' 'Chalk is white.' In these two sentences, is the word 'is' used (i) correctly? (ii) with the same meaning? The first question can be answered immediately, but to answer the second question we have to reflect on our use of the word 'is' in each sentence.

In these three examples the contrast is between two modes of functioning of intelligence: the intuitive and the reflective. At the intuitive level we are aware through our receptors (particularly vision and hearing) of data from the external environment, this data being automatically classified and related to other data by the conceptual structures described in Chapters 1 and 2. We also act on the external environment by the use of our voluntary muscles acting on our skeleton (a description which includes speech and writing). This activity is largely controlled and directed by feed-back of further information about its progress and result, again via our external receptors. In many cases, it can be entirely successful without any awareness of the intervening mental processes involved, for example, when reading aloud, driving a car or answering the question '6 + 5?'

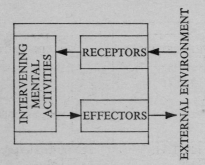

At the reflective level these intervening mental activities become the object of introspective awareness. A child asks us why we pronounce the word 'accelerate' as 'axelerate' not 'ackelerate'. So we explain (in terms appropriate to the hearer and with examples) that the first c is hard, because followed by a consonant, the second soft, because followed by e or i. Our pronunciation is explained by showing it to be consistent with certain accepted classes of response. Or, a learner-driver asks us why we changed gear before reaching the sharp bend in the road. Though we had done so 'without thinking' (that is to say without reflection), we have no difficulty in explaining our reason. Or, having replied '400' to the question '16 × 25', we might be asked 'How did you do that in your head?' And having described our method (there are several to choose from), we might also be asked to justify it – a much more searching question, as the answer involves reference to the associative property of multiplication.*

The data necessary to answer all of these questions comes not from the environment but from our own conceptual systems. These are represented in the diagram on page 53 by the block labelled 'intervening mental

* See Chapter 8, page 159.

activities'. We direct our attention to this source of data so easily and habitually that we take for granted this ability to reflect on our own mental processes. But we should be much more surprised at it than we are. Our awareness of the outside world can be accounted for by obvious sense organs – eyes, ears, etc. – and the neural paths from these are traceable. But no neuroanatomist has yet revealed the internal equivalent by which we can 'see' our own visual imagery or 'hear' our internalized speech.

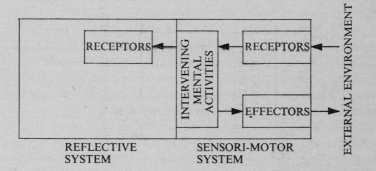

Moreover, this ability is much less developed in young children. Here are two examples from the work of Piaget: [1]

Weng (age 7): 'This table is 4 metres long. This one is three times as long. How many metres long is it?' *'12 metres.'* 'How did you do that?' *'I added 2 and 2 and 2 and 2 and 2, always 2.'* 'Why 2?' *'So as not to take another number.'*

Gath (age 7): 'You are 3 little boys and are given 9 apples. How many will you each have?' *'3 each.'* 'How did you do that?' *'I tried to think.'* 'What?' *'I tried to think in my head.'* 'What did you say in your head?' *'I counted . . . I tried to see how much it was and I found 3.'*

Being able to do something is one thing; knowing how one does it is quite another. There are, however, considerable individual differences in this, and the writer recently obtained the following replies from a child aged six years ten months. (To the first question, with feet instead of metres.) *'12 feet.'* 'Can you say how you found the answer?' *'Well I went 3, 6, 9, 12.'* (To the second question.) *'Three.'* 'How did you find out?' *'3 and 3 and 3 make 9.'* (Then, a spontaneous afterthought.) *'The quickest way is to write [sic] 3 times 3 is 9.'*

Once we have become able to reflect, to some degree, on our own schemas and how we use them, important further steps can be taken. We can communicate these, as in the foregoing example. We can set up new schemas and make new plans based on these. Someone unable to do the example cited earlier (16 × 25) might, after it had been pointed out that four twenty-fives make a hundred, not only be able to work out 16 × 25 by thinking of it as 4 × (4 × 25) which is equal to 4 × 100, but also work out other multiplications, like 24 × 25 and even 25 × 25. If the person could do all these, it would indicate that a simple schema and not just the answer to a particular question had been acquired.

We can replace old schemas by new ones. If readers have tried to back a car with a boat trailer or caravan attached, they may appreciate the following non-mathematical example. The writer had been told to 'put the steering wheel down on the side you want the trailer to go'. This was not very successful, however, and his co-driver suggested the following alternative approach. 'If you were just pushing it by hand, you would have no trouble in steering it, would you? So imagine yourself pushing on the towing hitch, but using the car to push with.' Substitution of this schema proved strikingly successful, since backing the car itself in any desired way was already automatic.

We can correct errors in existing schemas. If we say 'I see what I was doing wrong,' this implies not only a reflection on our existing method but the discovery of the particular details in it which were causing failure, followed, usually, by a deliberate change in these details.

Just how we are able to make deliberate changes in our schemas, as a whole or in detail, is still unknown. Since, however, we can certainly do so, our diagram needs further additions.

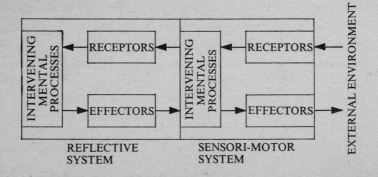

Here are some further examples which involve reflective activity.

Someone wants to know how to multiply together two decimal fractions, say, 1·2 and 0·57. So we explain to them how the decimal point may be omitted, the multiplication done in the ordinary way and the decimal point then re-inserted by counting the total number of figures after the decimal point. ($12 \times 57 = 684$; 1·2 has one figure after the decimal point, 0·57 has two – total three; re-insert the decimal point in the result to give three figures after the decimal point – result 0·684.) This rule will enable them to get the right answer, but it is unrelated to their existing knowledge of the meaning of decimal notation. To explain the method, we could rewrite the decimals as common fractions:

$$1·2 \times 0·57 = \frac{12}{10} \times \frac{57}{100} = \frac{684}{1000} = 0·684$$

The power of 10 in the denominator = the number of zeros in the denominator = the number of places after the decimal point. Multiplying the denominators corresponds to adding the numbers of zeros, which corresponds to adding the numbers of decimal places.

Having done all this, we could go further and reflect on our method of communication. We might then decide that it would be better next time to demonstrate the meaningful method first, before showing (or encouraging the learner to seek) the short cut. So we would reorganize our plan for communicating the schemas for multiplying decimals.

A far-reaching kind of reflective activity is that which leads to mathematical generalization. In the process of learning the use of indices, for example, we can distinguish two distinct stages. After defining the notation,* by examples such as

$$a^2 = a \times a \qquad \text{(where } a \text{ is any number)}$$
$$a^3 = a \times a \times a$$
$$a^4 = a \times a \times a \times a, \qquad \text{etc.}$$

it is easily seen that

$$a^2 \times a^3 = a \times a \quad \times \quad a \times a \times a$$
$$= a^5$$

* The reader to whom this notation is unfamiliar will find it further explained in Chapter 12, page 216.

and from this and similar examples learners form, intuitively, the general schema whereby they can write directly

$$a^5 \times a^7 = a^{12}, \qquad \text{etc.}$$

By using the methods for manipulating algebraic fractions already known to them, they can also form a schema for division, derived from examples such as

$$a^5 \div a^2 = \frac{a \times a \times a \times a \times a}{a \times a} = a \times a \times a = a^3$$

whereby they can write directly

$$a^{15} \div a^6 = a^9, \qquad \text{etc.}$$

Having formed these two (related) schemas, they can also *formulate* them: that is, express them symbolically in the form

$$a^m \times a^n = a^{m+n}$$
$$a^m \div a^n = a^{m-n}$$

where m and n stand for any two natural * numbers other than zero, and in the second case m is greater than n. These formulations detach the methods from any particular example of their use and enable them to be examined as entities in themselves. The restrictions that m and n must be natural numbers and m greater than n were made necessary by the initial definition of a^2, a^3 . . ., since symbols like a^0, a^{-2}, $a^{\frac{1}{2}}$, have no meaning in terms of this definition. But the methods have now become partially detached from their origins, and restrictions which at first seemed right and proper now become open to question. Under what conditions is it (1) permissible and (2) advantageous to remove these restrictions?

A reasonable criterion for the first is that the new method shall not introduce any inconsistency with known methods; and for the second, that by removing the original restrictions the advantages of index notations can be usefully and meaningfully extended.

Many readers will be familiar † with the extensions of index notation whereby

* That is, counting-numbers such as 1, 2, 3, etc.

† If not, please read Chapter 11, pages 217–20, before continuing. The mathematical argument was put there in order not to digress from the psychological argument here.

a° is given the meaning 1

a^{-2} is given the meaning $\dfrac{1}{a^2}$

$a^{\frac{1}{2}}$ is given the meaning $\sqrt{a}$

and so on. With these and similar meanings for negative and fractional indices, the original restrictions can be removed. We say that the notation and method have been *generalized*.

What are the mental processes involved?

From a set of examples, a general method is derived,

which can be applied to other examples *of the same kind*.

The method is next formulated explicitly, considered as an entity in itself, and its structure analysed.

This structure is used to invent ways of using the same method for examples of a new kind. The original examples are included in the enlarged field of application of the method.

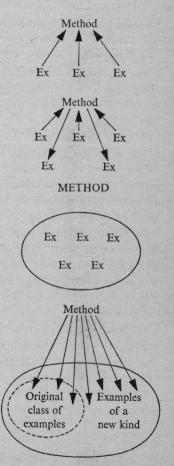

This process of mathematical generalization described above is a sophisticated and powerful activity. Sophisticated, because it involves reflecting on the *form* of the method while temporarily ignoring its content. Powerful, because it makes possible conscious, controlled and accurate reconstruction of one's existing schemas – not only in response to the demands for assimilation of new situations as they are encountered *but ahead of these demands*, seeking or creating new examples to fit the enlarged concept. Trying to do this intuitively is a much more hit-or-miss affair and will not perform to order. It must be accepted that the intuitive leap is a frequent forerunner of the deliberate generalization, suggesting a direction which might otherwise have remained unexplored. But intuition sometimes 'lets one down'. That is, when subjected to critical analysis, weaknesses are found – inconsistencies with accepted ideas, which make true assimilation to existing (and well-tried) principles impossible. The learned professor mentioned at the beginning of the chapter was right to be cautious of his intuitive judgement until he had tested it analytically.

A widely encountered example of successive mathematical generalization is that of number. Both historically and for the individual learner, the natural numbers come first. These are properties of sets of discrete (and so countable) objects, and methods for adding and subtracting, multiplying and dividing these, developed over the centuries, are learnt in their first decade or so by children of our own culture. Subsequently other things are encountered called 'fractions' and 'negative numbers', and rules are given which are alleged to be the correct way to add and subtract, multiply and divide, these. Correct according to what criterion? All too frequently the only one offered is whether the teacher decides that the rules have or have not been correctly followed. In such cases understanding has been lost, perhaps never to be recaptured. Worse, 'making sense' has ceased to be the criterion by which a mathematical statement is judged. Worst, another learner has been convinced that mathematics is boring and meaningless – true of what is presented under this guise, but false of mathematics.

How may the idea of number be successfully generalized through the stages of fractional numbers, integers, rational numbers, etc.? A detailed answer is given in Chapters 9 and 10, but it is worth taking a preliminary look at it here. Briefly, we need to formulate what are the formal properties of the system of natural numbers. By the *system* of natural numbers we mean the set of natural (counting) numbers together with the operations of addition and multiplication, whereby any two members of the set can be combined (in one way or the other) to get another member of the set. By the formal properties we mean those properties which do not depend on the particular examples we choose. So $12 + 9 = 21$ and $12 \times 9 = 108$ are not formal properties, but $12 + 9 = 9 + 12$ and $12 \times 9 = 9 \times 12$ are, though not stated in a general way. The five formal properties of the natural-number system are:

$$a + b = b + a$$
$$a \times b = b \times a$$
$$a + (b + c) = (a + b) + c$$
$$a \times (b \times c) = (a \times b) \times c$$
$$a \times (b + c) = a \times b + a \times c$$

where a, b, c, are any (natural) numbers. It is tempting to regard these properties as trivial, but they are the very foundation of all numerical manipulation, as is explained in Chapter 8. For example, without the first property the size of our shopping bill would depend on which we bought first; and without the third, it would depend on which two items the assistant added together first. These five properties are also, with the help of index notation, the foundation of elementary algebra, as is explained in Chapter 11.

Invaluable though our system of counting numbers is, it has its limitations. With the help of units it can be extended to make possible the measurement of continuous objects, but we soon find that the existing numbers do not include all we need to deal with sizes less than a unit. So new numbers corresponding to these broken units are introduced. But we are premature in calling them numbers – before we generalize the 'number system' schema, we must satisfy the two requirements of consistency and usefulness. (A pure mathematician would be content with the first, but the second usually follows, sooner or later. Few mathematical ideas of great generality fail to be useful in the course of time.)

Consistency means that we have to invent ways of 'adding' and 'multiplying' these new entities which have the five formal properties already listed. Usefulness means that the results of these manipulations must tell us something we want to know in terms of the material objects to which these entities refer. And though this is not an essential, it will be a great help if the signs for these new entities can be developed out of signs in common use already (just as we use the letters of our existing alphabet for newly invented words), and if the methods for 'adding' and 'multiplying' can make use of the large number of addition and multiplication results which we have already learnt. All these requirements, when satisfied, make possible the assimilation of the new number system to our existing and well-practised schema.

The way in which they are all met is a fascinating subject, and the reader who explores it further will learn much about the foundations of mathematical thinking. The same applies to the development of positive and negative integers, rational numbers (often identified with fractional numbers) and real numbers (the system which includes irrationals like $\sqrt{2}$, π). Here we are concerned mainly with the process rather than the result, and particularly with the activity of reflecting on the formal properties of the schemas, which is part of the process of mathematical generalization and one of the most advanced activities of reflective intelligence.

If this second-order functioning of intelligence is of such importance for progress to the more advanced levels of mathematics, it is clearly of great interest to know at what ages it makes its appearance, and how (if at all) we can aid and perhaps hasten its appearance. In answer to the first question, we have a body of research by Inhelder and Piaget [2] which indicates that children develop the ability to reflect on *content* during the ages from about seven to eleven and to manipulate concrete ideas in various ways, such as reversing (in imagination) an action, thereby returning to the previous state of affairs. They found, however, that their subjects could not reason formally – consider the form of an argument independently of its content – until adolescence. Closely related to this, they found that the younger children could not argue from a hypothesis if this hypothesis was contrary to their experience.

In this research the subjects were taken 'as they were found'. That is to say, the experiments indicate the progress of reflective intelligence as it developed in the Swiss schoolchildren from whom the subjects were taken, by the interaction of their innate abilities with the cultural and educational experiences which they encountered. What we do not at present know is the extent to which the rate of development might be helped by teaching. To consider a parallel: most children learn to sing spontaneously, just from growing up in a culture where they hear other people singing. But this learning is greatly accelerated, and the final degree of accomplishment greatly raised, in, say, boys who become members of the choir of King's College, Cambridge, or Magdalen College, Oxford. At present the development of reflective ability and formal reasoning is not the subject of deliberate teaching, partly because its importance has hardly been realized and partly because we do not know how to teach it, which presupposes a knowledge of how it is learnt.

A reasonable hypothesis about the latter is that any situation which requires learners to formulate ideas explicitly, and to justify them by showing them to be logically derivable from other and generally accepted ideas, would exercise and so develop the ability to reflect on their schemas. In other words, argument and discussion are useful ways of learning to reflect.

Those who have tried usually agree that trying to teach a topic exerts strong pressure to clarify one's own thinking about it. A simple experiment has given support to this view. Two parallel forms of secondary-school boys, aged about fourteen, were taught different topics by their regular mathematics teachers. Each form was given a test on the topic it had been taught and divided into two halves of equal performance (as nearly as possible) as measured by these tests. One half of each form then taught what they had learnt to their opposite numbers in the other form, while the other (matched) half spent the same time in further practice at the same topic. The boys who were acting as teachers thought that their pupils were going to be tested on what they had been taught by them. Actually, at the end of the experiment all were re-tested on the topic which they had learnt during the first part of the experiment, the aim being to compare the effects of teaching a topic to someone else and continuing to practise it by oneself. The results came out quite clearly in favour of the former.

Communication seems to emerge as one of the influences favourable to the development of reflective intelligence. One of the factors involved is certainly the necessity to link ideas with symbols: this will be considered at length in the next chapter. Another is the interaction of one's own ideas with those of other people. To the extent that agreement is reached, the resulting communality of ideas is less egocentric, more independent of individual experience. And as has already been suggested, the cut and thrust of intellectual discussion forces on one the necessity to clarify ideas in one's own mind, to state them in terms not likely to be misunderstood, to justify them by revealing their relationships with other ideas; and also, to modify them where weaknesses are found by the other side, ending with a stronger and more cohesive structure than before. To be able to do the last requires the achievement of a partial detachment from one's ideas, a state of reduced involvement, so that one does not feel personally attached, injured or defeated when one's schemas are shown to have some inaccuracy or inconsistency. This is yet another aspect of the state of reflection. It is also largely dependent on the interpersonal situation, an aspect which will be discussed in Chapter 6. The last consideration suggests that relationships with teachers may be of great long-term importance in the development of reflective intelligence. But on this, there is as yet no evidence.

A caveat is desirable here. The preceding discussion has to some extent carried the implication that an individual is 'at the intuitive stage', 'capable of reflection on form and content combined', 'capable of formal reasoning', in general; that is, if he is at a given stage relative to topic A, he is at the same stage relative to topic B. But it may well be the case that we all have to go, perhaps more rapidly than the growing child, through similar stages in each new topic which we encounter – that the mode of thinking available is partly a function of the degree to which the concepts have been developed in the primary system. One can hardly be expected to reflect on concepts which have not yet been formed, however well-developed one's reflective system. So the 'intuitive before reflective' order may be partially true for each new field of mathematical study. Further research is needed here.

While learners are still at the intuitive stage they are largely dependent on the way material is presented to them. If the new concepts encountered are too far removed from any of their existing schemas, they may be unable to assimilate them, particularly if reconstruction is required, for this is largely dependent on reflection. So in the earlier stages a conceptual analysis by the teacher must be used as a basis for a careful plan of presentation, from which learners can re-synthesize the structures in their own mind. This is the case whether learning takes place directly from a live teacher, or indirectly from a book. The former situation has the advantage that questions can be asked, explanations given; an even greater advantage is that a sensitive teacher can perceive the growing points of the learner's schema and offer the right material at the right moment. This flexibility of approach entails also a greater mastery of the subject than keeping strictly to a prepared plan, however good.

The final contribution of the excellent teacher is, however, gradually to reduce the learner's dependence. When my young son was first doing jigsaw puzzles, his mother or I used to offer him pieces which fitted on to what he had already put together. But a stage was reached when he did not like us to do this any more, and it is towards this kind of independence that the mathematics teacher must work. Once people are able to analyse new material for themselves, they can fit it on to their own schemas in the ways most meaningful to themselves, which may or may not be the same ways as it was presented.

So the teacher of mathematics has a triple task: to fit the mathematical material to the state of development of the learners' mathematical schemas; to also fit the manner of presentation to the modes of thinking (intuitive and concrete reasoning only, or intuitive, concrete reasoning and also formal thinking) of which the pupils are capable; and, finally, to increase gradually the pupils' analytic abilities to the stage at which they no longer depend on their teacher to predigest the material for them.

And although we have some reasonable conjectures about how this last development may be encouraged, our knowledge in this area is far from complete. In this respect, as in many others, the best teachers are those who are still active learners.

CHAPTER 4

Symbols

In previous chapters we have considered the formation of concepts, the function of schemas (conceptual structures) in integrating existing knowledge and assimilating new knowledge, and the additional power which comes from the ability to reflect on one's schemas. In each of these processes an essential part is played by symbols, which have other functions as well. It is now time to consider these in detail.

Among the functions of symbols, we can distinguish:

(i) Communication
(ii) Recording knowledge
(iii) The communication of new concepts
(iv) Making multiple classification straightforward
(v) Explanations
(vi) Making possible reflective activity
(vii) Helping to show structure
(viii) Making routine manipulations automatic
(ix) Recovering information and understanding
(x) Creative mental activity

Most of these are related, particularly to the first. Recording knowledge is communicating with the reader, explanation is a special kind of communication, reflecting is communicating within oneself; and other connections will also be apparent. The headings are therefore intended for convenience only, as starting points for the discussions which follow, not as partitions.

(i) *Communication*

A concept is a purely mental object – inaudible and invisible. Since we have no way of observing directly the contents of someone else's mind, nor of allowing others access to one's own, we have to use means which are either audible or visible – spoken words or other sounds, written words or other marks on paper (notations). A symbol is a sound, or something visible, mentally connected to an idea. This idea is the *meaning* of the symbol. Without an idea attached, a symbol is empty, meaningless.

Provided that a symbol is connected to the same concept in the minds of two people, then by uttering* this symbol, one can evoke the concept from the other's memory into their consciousness – can cause them to 'think of' this concept in the present. This proviso is, however, no small one. Once the connection is established, its meaning is projected on to the symbol, and the two are perceived as a unity. So it is hard to realize that what is meaningful to oneself may not be meaningful to the hearer – a difficulty experienced by many when speaking to foreigners – or that the same meaning is not being attached, for example, to the word 'braces', which may mean to someone British a device for holding up one's trousers, but to an American a pair of set brackets { }. We may think that we are communicating when we are not, and, indeed, it is impossible to know for certain whether we are, and, if so, to what degree. For the reason given above, we usually take it for granted, but the communication links are so precarious, and so inaccessible to study, that we would do better to be surprised that we can communicate our ideas to each other at all. After all, it has taken millions of years of evolution to produce an animal which can do so to any marked extent.

Let us take as a starting point (a) that a symbol and the associated concept are two different things; (b) that this distinction is non-trivial, being that between an object and the name of that object. If an object is called by another name, we do not change the object itself, and this is still true for an object of thought – in the present context, a mathematical idea. For example,

$$\text{'five', 'cinq', '5', 'V', '101'}$$

all refer to the same number in different notations. We do not call five an English number and *cinq* a French number, nor should we call 5 an Arabic number and *V* a Roman number. But we still read, all too often, instructions to pupils like 'Turn the binary number 101 into a decimal

* This will be used as a convenient condensation for speaking, writing, drawing, projecting on a screen, etc.

number.' The whole object is, of course, *not* to change the number itself in the process of representing it in a different way. In translating from French into English, we try to keep the meaning the same while changing the words. In converting pounds to dollars we try to keep the value in goods or services the same while representing this value by different tokens (coins, notes) or symbols (figures on a cheque or bank transfer).

The term 'binary number' also implies that being binary is a property which a number can have or not, like being even, prime, an integer, etc. But binary *numerals* can be used to represent any kind of number at all, odd or even, prime or factorizable, natural number, integer, rational or real number. One of the first requirements of communicating an idea is to be clear about it oneself. Those who talk or write about 'binary numbers' and 'decimal numbers' are not.

Usually, when uttering a symbol, we want to call to the attention of the receiver the idea attached to the symbol rather than the symbol itself. If it is the symbol we are referring to, we can show this by quotation marks. (More symbols! They are inescapable.) Example:

'5' and 'V' are both symbols for (the number) five.

A symbol for a number is called a 'numeral', and a system of numeration is a system for writing as many different numbers as we like with a relatively small number of digits (single numerals like 0, 1, 2, 3, 4 . . . 9). The decimal system uses ten digits, the binary system uses two. If it is not clear from the context which system is in use, this can be shown simply and clearly by a suffix. The sign = ('is equal to') means that we are referring to the same concept, (usually) by different symbols. So, for example,

$$5_{ten} = 101_{two} \text{ (since 101 in binary means the same as 5 in decimal notation.)}$$

Similarly $8_{ten} = 10_{eight} = 1000_{two}$ etc.

But   '8_{ten}' $\neq$ '10_{eight}'. The numbers are equal, the numerals are different.

Excessive precision in the use of language* is rightly regarded as pedantry. So it is a fair question at this stage to ask whether this label is

---

* A symbol system; for example, the English language, the language of mathematics.

applicable to the preceding discussion. Does it really matter, for example, which of these we say or write:

'Write the binary number 11010 as a decimal number' or
'Write 11010_{two} in decimal notation'?

An easy defence would be to claim that it is part of the duty of a mathematician to be as accurate as possible all the time. But this, though plausible, is not valid. It would, for example, imply that we should never use convenient but loose phrases such as 'as small as we like'. Part of the aim of mathematics is, by abstraction and the omission of irrelevancies, to enable us 'to see the wood for the trees', and this will not be achieved by adding, instead, a mass of mathematical detail in the name of accuracy.

The kind of accuracy with which we are at present concerned is accuracy of communication, with trying to get as near as we can to the impossibility of producing the same idea in the mind of the receivers as of the communicators or calling it to their attention.

Now, we can distinguish three categories of hearer or reader. First, those who don't yet know what we are talking about, but want to. For these, we should choose our symbols with the greatest possible care and use them as accurately as we can, with the aim of communicating nothing but the truth, though not yet necessarily the whole truth. Concepts are built up by degrees. The first approximation is bound to be incomplete, and perhaps to need tidying up in detail, but there should not be anything of importance to un-learn. It is also worth bearing in mind that, to an intelligent learner, a brief but inaccurate statement may well be more confusing than a somewhat lengthier, but accurate, statement.

The second category comprises those who do know what we are talking about, as a general background within which we are trying to communicate some particular aspect. If they are willing to 'go along' with us, we can take much for granted, save time and concentrate on essentials. An old and wise teacher of mine often used, in the context of limits and convergence, phrases like 'As near as dammit to . . .' We both knew what he meant, and both could, if necessary, have re-phrased it in rigorous terms. So, for the task in hand, the idea was communicated with complete accuracy by this short and expressive phrase.

The third category of hearer or reader consists of those who do know what we are talking about but want to fault it. A non-mathematical example of this activity is to be found every time a new tax is made law. The finance minister says 'I want a tax on . . .' As soon as this becomes law, an army of expert accountants will go to work on behalf of their clients to see how this tax can legally be avoided or reduced. So, before the bill goes through, the parliamentary draughtsmen have to try to stop all loopholes in advance. The result is to make it almost unintelligible.

Similarly in mathematics, rigour and ease of understanding do not go together. The art of communication is, first, to convey meaning. Afterwards, the new ideas can be subjected to the stress of analysis, and greater precision introduced where weaknesses are found. The difference is that, once a schema is well established, this critical attack serves a useful purpose, that of stimulating more careful formulation and greater reflective awareness, and the strengthening of the schema without loss of integration of 'the overall picture'. This criticism may come from another person or it may come from a 'devil's advocate' within oneself. This seems to be another function of the reflective system – to take 'an outside view' of an argument or other intended communication, and by self-criticism anticipate external criticism.

(ii) *Recording knowledge*

Ideas are not only invisible and inaudible, they are perishable. When we die, our knowledge dies with us, unless we have communicated or recorded it. One of the most moving episodes in the history of mathematics is that of the young Galois (1811–32) sitting up all night, writing against time to commit to paper his theory of groups, before his tragic and wasteful death by duel at the age of twenty.

Recording is a special case of communicating, since it is normally done with the intention that these records will, in the near or distant future, be seen by others. So all the previous section applies. Whereas the spoken communication usually (though not always) takes place in circumstances which allow questions and explanations to be given, written or printed symbols have to convey all the required meaning, without a second chance on either side. So the communicators have to take more trouble to try to ensure this. There is, however, the advantage that the receivers have a permanent record for revision and the checking of earlier points. They can also go at a speed to suit their own rate of assimilation.

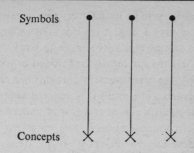

As has been discussed in Chaper 1, the conceptual structure of mathematics is something far beyond that which anyone could construct, unaided, in a lifetime. Limited areas have taken years of work by some of the world's most gifted individuals. It is the storage of the accumulated knowledge of previous generations by written and printed symbol systems (and recently by other techniques such as recording tape, cinematography, microfilm), together with the auxiliary explanations of live teachers, that make it possible for some of each new generation to learn in years ideas which took centuries of collective effort to form for the first time, in each case, synthesizing them anew, and in some cases building new knowledge and adding this to the store.

One of the first requirements for the avoidance of ambiguity which one would expect to be observed is that each symbol is associated with one concept, and vice versa. This arrangement is, however, seldom found in practice, even in a single language. Mathematicians seem to be particularly lazy about inventing new symbols, relying largely on the capital and lower-case letters of the Roman alphabet, the Greek alphabet,

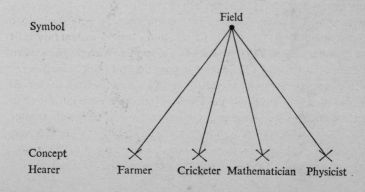

punctuation marks and the like, each of which does multiple duty. So a single symbol may well stand for a variety of concepts.

The arrangement at the foot of page 69 might be expected to lead to confusion, since the word 'field' will evoke different concepts in the minds of each of the individuals named above. Or, if we are addressing someone with interests in all these topics, then we cannot be sure which concept will be evoked by the word 'field' in isolation.

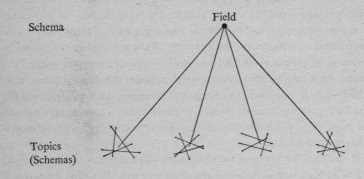

Schema — Field

Topics (Schemas)

But, of course, the word is seldom used in isolation. Ordinarily the hearer knows which topic is under discussion, and only ideas within this topic are accepted as possible meanings for the word. If not, then the speaker or writer uses one or more other symbols to evoke the relevant schema as a whole. This establishes a 'set' – a state of mind in which concepts belonging to this particular schema are more easily evoked. Symbols used in this way, to determine the schema within which a particular symbol takes its meaning, are called its *context*.

From this, three simple rules can be formulated for conveying the desired meaning when one symbol corresponds to many concepts.

(1) Be sure that the schema in use is known to the hearer or reader.

(2) Within this schema let each symbol represent only one idea.

(3) Do not change schemas without the knowledge of the hearer or reader.

It is permissible (though whether any advantage is gained is another question) to use the same symbol in different contexts with different meanings. But in the same context a symbol must have just one meaning.

So we can write $AA' = I$ in the context of matrices, and $AA' = BB$ in the context of points and lines, without confusion. But if we write $(x + a)^2 = x^2 + 2ax + a^2$ the x and the a must keep the same meaning throughout, because they are in a single context.

These rules seem straightforward and obvious, but they are not always observed, with the result that the learner is confused. Here is an example.

Children first learn the meaning of multiplying in the context of natural numbers, which refer to sets of discrete, countable objects. So the operation 3×4 corresponds to combining four sets, each of three objects, and counting the objects in the resulting set.* They use the sign '$\times$' with this meaning for several years, and it is the only meaning they know. We then change to a new number system, say, fractional numbers or integers, in which the sign (or word) has a different meaning. But we do not tell the children that we have changed the context and have generalized the meaning of '$\times$' to suit the new context. So they no longer fully understand what they are doing.

If the new context was very different from the old, children would probably discover what was happening unaided. But the contexts are sufficiently alike to make it hard for them to do so. One way of indicating the change is already in use in advanced texts. The symbol '$\otimes$' (and also '$\oplus$') is used in the new context, to show that these operations are like the others but that we must not expect them to be quite the same. The readers of these texts probably come into the third of the categories outlined on pages 67–8, those who will be quick to notice any inaccuracy. But those for whom accuracy of communication is most necessary are those in the first category, those who do not yet know what we are talking about but want to. When these pass on into category two, we may conveniently revert to the symbols '$+$' and '$\times$', since they are now able to assign the appropriate meanings according to context.

The word 'line' is commonly used with at least three different meanings: (a) a line of indefinite length, extending indefinitely in both directions; (b) one which starts at a given point and extends indefinitely in one

---

* That is, assuming that we read '3×4' as 'three multiplied by four'. It is also read by some as 'three times four', which corresponds to combining three sets, each of four objects. We should be more surprised than we are that both of these give the same result.

direction from it; and one which is of finite length, bounded by two points. These three meanings may conveniently be distinguished by the terms 'line', 'ray' and 'line segment'. So the point X is on the line AB (or BA), and also on the ray BA; but it is not on the ray AB, nor on the line segment AB. If AB represents a railway line, X our destination and A our starting point, the distinction is hardly trivial!

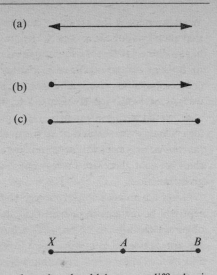

The mathematically experienced reader should have no difficulty in finding other examples of the ambiguous use of symbols. Some suggestions: what is meant by '$AB = 3$ cm.'? What is meant by 'the series $1 + \frac{1}{2} + \frac{1}{4} + \frac{1}{8} + \frac{1}{16}$ etc.'? And in the context of groups, are the terms 'identity element' and 'neutral element' synonymous?

So far, the emphasis of this section has been that, in a given context (which may be explicit or implicit), one symbol should represent only one concept.

We must have this,

not this.

But, perhaps surprisingly, this is even better.

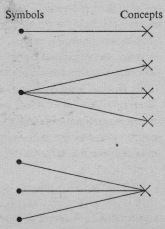

What matters is the meaning (the associated concept), and provided that each symbol conveys only one meaning, it is often an advantage to have a choice. If A uses a term (for example, 'cuboid') which is unfamiliar to B, they can try again with another (say, 'rectangular block'). The choice of symbol also enables us to classify the same idea in different ways, a use which will be discussed further in section (iv) of this chapter; and, related to this, it can help us to emphasize that aspect of a complex idea which is most relevant to particular circumstances. For example, *function* is a concept with widespread applications, and in Chapter 13 we shall see that there are no less than six useful ways of representing a given function.

Other advantages of using several different symbols for the same concept will be mentioned later in this chapter. If we do this, however, an obvious precaution is necessary to ensure that the reader knows that we are in fact talking about the same thing, though using different names; and this becomes more important when recording mathematics, as distinct from communicating face-to-face, since the reader cannot ask. This is the meaning of the symbol '=', that the symbols on each side of the sign of equality refer to the same object.

(iii) *The communication of new concepts*

It will be recalled that in Chapter 1 the point was made that new concepts of a higher order than those which the learner already has can only be communicated by arranging for the learner to group together mentally a suitable set of examples.

If the new concept is a primary concept, for example, red, it is possible to do this without the use of symbols, simply by pointing. The words 'This is a . . .' simply help to draw attention; they are verbal pointers. 'Red tie', 'red book', 'red pencil', 'red light', however, express simultaneously the variability of the examples and the constancy of the concept. Intuitively the learner associates the invariant property with the invariant word, and so learns the name for the concept while it is being formed.

If the concept is a secondary concept, as are all mathematical concepts, then the only way of bringing together a suitable set of examples in the learner's mind is to bring together the corresponding words. 'Red, blue, green, yellow – these are all colours.' By manipulating the words we manipulate the minds of the learners – normally, with their consent. (If they feel otherwise, there will naturally be resistance to learning: see Chapter 6.) In this way learners may be helped to see something in common between examples which, separately encountered over an interval of time, would have remained isolated in their minds. It took Newton to perceive for the first time something in common between the fall of an apple and the motion of the planets round the sun, but when he brings these ideas together for us, we too can form the concept of gravitation.

Another way of communicating new concepts is by relating together classes already known to the hearer. 'What is a Sinhalese?' 'An inhabitant of Sri Lanka.' 'What is a kite?' (In the context of geometry.) 'A quadrilateral with two pairs of adjacent sides equal.' 'What is a variable?' 'An unspecified member of a given set.' If the hearer already has the class concepts mentioned, this implies that examples of these are known, so it should also be possible to supply examples of these new concepts. Indeed, this is often the first response, partly to confirm that the concept has been understood. (Sketching rapidly in response to the second definition: 'Like this?')

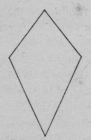

But the response also seems to satisfy a deeper need. Somehow, a concept acquired in the way just described seems incomplete until it has some examples. A tentative explanation of this is that a concept confers the ability to class together an appropriate set of examples, and it is generally observable that the acquisition of a new ability often seems to carry with it a need to exercise it. (Give your small son a kit of tools for his birthday and observe the result.)

The examples of the new concept thus supplied need not be from past experience. One can imagine a Sinhalese without ever having met one;

one can imagine a 100-sided regular polygon without having seen one and without having to draw one. Indeed, a fruitful and exciting method of mathematical generalization is to invent a new class, and then try to find some members of it. Example: suppose that we already have the concepts square root and negative number, and combine these to form a new concept – the square root of a negative number. The search for examples of this new class, and the investigation of their properties, leads to the construction of a new set of ideas which, though termed 'imaginary' numbers, are nevertheless of great practical use in physics: for example, in the theory of alternating current and oscillatory circuits.

(iv) *Making multiple classification straightforward*

A single object may be classified in many different ways, and, by using different names for it (which we have already seen to be permissible), we can indicate what particular classification is currently in use. The same man may be called 'Mr John Brown', 'Sir', 'The right honourable gentleman', 'Uncle Jack,' 'Daddy', or 'John'. The same angle may be classified as the angle vertically opposite to . . . or as the third angle of triangle . . . The same number may be regarded as the square of 8, the cube of 4 or the square of 10 minus the square of 6, may be symbolized by 8^2, 4^3, $10^2 - 6^2$. By our choice of symbol, we are enabled to concentrate our attention on different properties of the same object.

As already noted, we show that we are still (often in spite of appearances) referring to the same object * by the symbol '=', and, by renaming according to already established routines, we can find properties which were at first not apparent.

Example: $4x^2 - 12xy + 9y^2$, where x and y are both numerical variables (unspecified numbers). We know that this collection of symbols represents some number. But by writing

$$4x^2 - 12xy + 9y^2 = (2x - 3y)^2$$

we know something new – that it represents a *positive* number.

Though the principle is a simple one, its consequences are far-reaching. Once we have appropriately classified something, we are a long way towards knowing how to deal with it. (This polite caller – is he a salesman,

* Reminder: this, in the present context, usually means an object of thought.

a public-opinion surveyor or a plain-clothes detective? Our response is cautious until we know which.) 'Appropriately' means in a way (or ways) which helps us to solve the problem in hand; and so the more ways in which we can classify, the greater the variety of problems which we can solve. And the more symbols we can attach to the same concept, the more ways can we classify.

(v) *Explanations*

An explanation is a communication intended to enable someone to understand something which they did not understand before. Understanding results from assimilation to an existing schema, so where this has failed there are three possible causes.

(a) The wrong schema may be in use. In this case the explanation needed is simply to activate the appropriate schema. In the present book, words such as 'function', 'image', 'group', are used both in the everyday sense and in the mathematical sense. Failure of understanding could result from attaching a different meaning from that intended. This is simply a matter of context.

(b) The gap between the new idea and the (appropriate) existing schema may be too great. Using again the indices example (page 55), suppose that one began by showing the notation

$$a^2 = a \times a$$
$$a^3 = a \times a \times a$$

and then continued straight to $a^m \times a^n = a^{m+n}$.
Very likely the learners would say that they did not understand, perhaps adding 'You have gone too fast.' The explanation needed here would be to supply more intervening steps, thereby bridging the gap. In psychological terms, the explainer would utter suitable symbols by which to evoke concepts relating the existing schema to the new idea.

(c) The existing schema may not be capable of assimilating the new idea without itself undergoing expansion or restructuring, of which a special case is mathematical generalization. In this case, the function (in the psychological sense) of an explanation is to help the subjects to reflect on their schemas, to detach them from their original sets of examples, which are now having a restrictive effect, and to modify them appropriately. The extension of index notation to zero and negative and fractional numbers would offer an example of this, if the new idea were presented in advance of the communications necessary to make it understandable. This seems to me to be a perfectly suitable way of teaching. It is not desirable never to put before the learner anything which does not relate by easy stages to what is already known. This 'over-programming' offers no challenge and no variety. It is often valuable to look first at the problem – say, that of finding the instantaneous velocity of a body in free fall; next, to analyse the tasks involved, which here include deciding on a suitable meaning for 'instantaneous velocity'; and methodically to develop, by the processes described, the new concepts (such as that of a limit) required to solve the problem.

(vi) *Making possible reflective activity*

This involves becoming aware of one's own concepts and schemas, perceiving their relationships and structure, then manipulating these in various ways. These three functions of the reflective system are represented in the diagram by the rectangles labelled 'receptors', 'intervening mental processes' and 'effectors'. In the present context these intervening processes are cognitive, and make possible the overall activity which we call reflective intelligence. These are not, however, the only intervening processes which may occur: another variety will be discussed in Chapter 6.

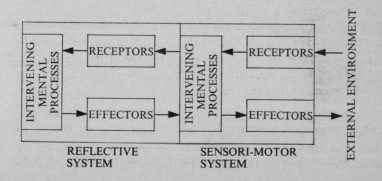

The process of becoming aware of one's concepts for the first time seems to be quite a difficult one. As was mentioned in Chapter 2, the overall development of this ability extends over a number of the years of childhood. But even in persons with a highly developed reflective ability, it is still a struggle to make newly formed, or forming, ideas conscious.

Making an idea conscious seems to be closely connected with associating it with a symbol. Just why this should be so is not yet known. Concepts are elusive and inaccessible objects, even to their possessors, and it may be that symbols (which are themselves primary concepts) are the most abstract kind of concept of which we can be clearly aware. Certainly more knowledge of the process would greatly increase the power of our thinking. Once the association has been formed, the symbol seems to act as a combined label and handle, whereby we can select (from our memory store) and manipulate our concepts at will. *It is largely by the use of symbols that we achieve voluntary control over our thoughts.*

Verbal thinking (which can be extended to include algebraic and any other pronounceable symbols) is internalized speech, as may be confirmed by watching the transitional stages in children. The use of pronounceable symbols for thinking is closely related to communication; one might describe it as communication with oneself. So becoming conscious of one's thoughts seems to be a short-circuiting of the process of hearing oneself tell them to someone else. This view is supported by the common observation that actually doing so to a patient listener (thinking aloud) is nearly always helpful when one is working on a problem. Visual thinking is a much more individual matter, and the relation between these two kinds of imagery will be discussed further in the next chapter.

(vii) *Helping to show structure*

This function of symbols is related to that of the preceding section, since one of the aims of reflection is to become aware of how one's ideas are related, and to integrate them further. But the span of immediate memory is small: that is, the amount of information which one can keep in consciousness at a time is very limited. Moreover, the more difficult the topic, the more one needs to concentrate one's attention on one thing at a time. But one also needs to be able to refer quickly and easily to previous work. So one records one's thoughts on paper as one progresses. This is a more permanent form of the 'thinking aloud' discussed in the previous section, which reduces the cognitive strain of keeping the whole of the relevant information accessible.

Another way of reducing the cognitive strain, and a powerful one, is the brevity of mathematical notation. Compare:

$(x + a)^2 = x^2 + 2ax + a^2$ The square of the sum of two numbers is equal to the sum of their squares plus twice their product.

Df the derivative of (the function) f.

$D^{-1}f$ the antiderivative of f.

2751   two thousand seven hundred and fifty one.

$\delta > O; EN$:   Given any positive number δ, there is a
$n \geqslant N, |x_n - x| < \delta$ number N such that for all values of n which are equal to or greater than N, the difference between the nth term of the sequence $x_1, x_2, x_3 \ldots$ and x is less than δ.

But there is more to it than this. A good or bad symbol system can be a great help, or a severe hindrance, in evoking and manipulating the right concepts in the right relationships. Here is one example. First, do a simple multiplication, say, 34 × 7, in the familiar place value notation. Now, repeat (if you can) in Roman numerals: XXXIV multiplied by VII.

Another example:

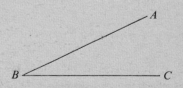

A conventional way to name this angle is ∠ ABC. This suggests that the vertex of the angle is at A, which it is not. If there is no ambiguity, we refer to ∠ B, not ∠ A. The alternative usage $\widehat{ABC}$ is better from this point of view, though less popular with printers. But let us now think about the ideas which we want to symbolize.

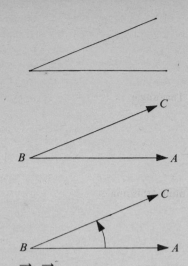

An angle is determined by
two directions through a
point. Each direction can
be represented by a ray
through that point, so an
angle is represented by two
rays, $\overrightarrow{BA}$ and $\overrightarrow{BC}$. If we are
concerned with a *turn*
from one direction to
another, represented by a
signed angle (the
convention is to take an
anticlockwise turn as
positive), this will be
represented by an ordered pair of rays $(\overrightarrow{BA}, \overrightarrow{BC})$, which can be condensed
to B_A^C and, where no ambiguity would result, to B.

Addition and subtraction of angles at a point can be reduced to
simple algebra once the 'rule of cancellation' has been seen and memorized.

$$B_A^C + B_C^D = B_A^D$$

Subtraction of an angle is
equivalent to adding its
inverse, that is, the angle
corresponding to a turn in
the opposite direction.

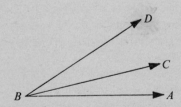

So
$$B_A^D - B_C^D = B_A^D + B_D^C$$
$$= B_A^C$$

What happens if we try to subtract a larger angle from a smaller? Let us
try.

Algebraically,
$$B_C^D - B_A^D = B_C^D + B_D^A$$
$$= B_C^A$$

To what does this correspond in the figure?

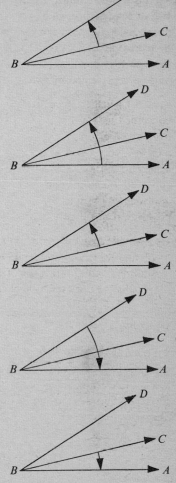

This angle

minus this angle

is equal to this angle

plus this angle

which is equal to this angle:
which makes sense in terms of
successive turns.

Reverting now to the first equation, $B_A^C + B_C^D = B_A^D$

and comparing it with $\overrightarrow{AC} + \overrightarrow{CD} = \overrightarrow{AD}$, the reader who is familiar with the addition of free vectors will recognize that the whole of the above algebra of (signed) angles is isomorphic with that for the addition and subtraction of free vectors, letters being read in the generally accepted positive directions of upwards and left to right.

There is no hope of replacing the long-established notation for angles by the above – the only hope of establishing a good notation is at the outset. Readers may not, in any case, be convinced by these arguments in favour of it. But it is hoped that they will at least agree that the choice of a notation should be made with care, and with some attention to how effectively it represents the ideas which (as this chapter has been devoted to showing) are so dependent on it.

The last example in this section was brought to my attention by Mr J. S. Friis. It requires an elementary knowledge of modular arithmetic and groups.

It is easily verified that $\{0, 1, 2, 3\}$ is a group under $\oplus$ mod 4, and so is $\{1, 2, 4, 3\}$ under $\otimes$ mod 5. The combination tables are, respectively,

$\oplus$ mod 4	0	1	2	3
0	0	1	2	3
1	1	2	3	0
2	2	3	0	1
3	3	0	1	2

$\otimes$ mod 5	1	2	4	3
1	1	2	4	3
2	2	4	3	1
4	4	3	1	2
3	3	1	2	4

It can be seen from these tables that these two groups are isomorphic, both being cyclic groups of order 4.

But if the right-hand table is rewritten in powers of 2, the isomorphism is revealed rather nicely.

$\otimes$ mod 5	2^0	2^1	2^2	2^3
2^0	2^0	2^1	2^2	2^3
2^1	2^1	2^2	2^3	2^0
2^2	2^2	2^3	2^0	2^1
2^3	2^3	2^0	2^1	2^2

The reader will also observe how smoothly index notation generalizes to the new context. This, surely, is another criterion for a good notation.

(viii) *Making routine manipulations automatic*

Thinking is hard work. Once we have understood a mathematical process, it is a great advantage if we can run through it on subsequent occasions without having to repeat every time (even though with greater fluency) the conceptual activities involved. If we are to make progress in mathematics it is, indeed, essential that the elementary processes become automatic, thus freeing our attention to concentrate on the new ideas which are being learnt – which, in their turn, must also become automatic. At any level, we can also distinguish between routine manipulations and problem-solving activity; and unless the former can be done with minimal attention, it is not possible to concentrate successfully on the difficulties. The same is true of any skill. To be a good driver, one must be able to change gear without thinking. Violinists cannot give themselves to the interpretation of the music until their technique is effortless.

In mathematics, this is done by detaching the symbols from their concepts and manipulating them according to well-formed habits without attention to their meaning. This automatic performance of routine tasks must be clearly distinguished from the mechanical manipulation of meaningless symbols, which is not mathematics.* Machines do not know what they are doing. Mathematicians, working automatically, can at any time they wish pause and re-attach meanings to the symbols; and they must be able to pass easily from one form of activity to the other, according to the requirements of the task.

The economy of effort involved is striking. First, we learn to manipulate concepts instead of real objects; then, having labelled the concepts, we manipulate the labels instead. (And if the manipulations can be reduced to a mechanical process, we can even program a computer to do them for us.) Finally, perhaps, we reverse the process by re-attaching the concepts to the symbols and then re-embodying the concepts in the real actions with real objects from which they were first abstracted. So we calculate, say, the stresses involved, and design (mechanical) structures to withstand these stresses, for an aeroplane to fly at twice the speed of sound – before it leaves the ground and even before the first plates are riveted together. The power of mathematics is immense, and at all stages symbols make a major contribution to this power. But without the ability of mathematicians to invest them with meaning, they are useless.

* We need to separate the two meanings, and the use of the two words automatic and mechanical seems a convenient way to do so.

(ix) *Recovering information and understanding*

This function of symbols is somewhat like those discussed in section (ii), the recording of knowledge, and section (vi), in which symbols were described as a combined label and handle for identifying and manipulating concepts. Here we are concerned with using them for bringing concepts and schemas back into availability from one's long-term memory store. Even concepts in current use are elusive objects; those which have not been employed for some time may be quite inaccessible without, so to speak, some kind of handle whereby to pull them back.

Try this experiment. Ask a man what shape a reflector must be to give a parallel beam. Does he reply 'It is a surface formed by revolving about an axis of symmetry a curve having the property that for some point S on the axis, and all points P on the curve, a ray SP and the line through P parallel to the axis are equally inclined to the tangent at P to the curve. This curve is called a parabola.' Or does he immediately reply 'a parabola', afterwards explaining the properties just described? In other words, does he first recall the conceptual structure, or does he first recall its label?

Another example, for those with some knowledge of quadratic equations. Has the following equation real roots? $3x^2 - 4x + 2 = 0$. In the majority of cases, either the word 'discriminant' or the symbols '$b^2 - 4ac$' will first come to mind. Afterwards, the method based on this will be recalled.

One more example. Have you ever met, say, old schoolfellows or colleagues and not recognized them; but as soon as they said 'I'm ...', you not only recognized but remembered much else about them?

This process of recovery of information by the help of symbols is exemplified by all mnemonics. One example will suffice.

This is a well-known device for remembering the signs of the trigonometrical ratios of angles from $0°$ to $360°$. The diagram gives the positive ratios.

The difference between a mnemonic and a formula is that the latter embodies the structure of what is to be recalled. From a formula, therefore, understanding can be reconstructed, even if it does not immediately follow the recall of the symbol. By hearing or seeing the words 'Ohm's law', the formula* $\frac{E}{I} = R$ will, for many persons, be evoked – that is, it will come into present consciousness from the long-term memory store. From further consideration of the formula it is easy to reconstruct its meaning: that for a given circuit the ratio of electromotive force to current is constant – double the volts and the amps will also be doubled.

This order of recall – symbols first, then meaning – is not, however, invariable. When recalling a conversation, or something read, most people reproduce the meaning but express it in their own words. Similarly, teachers sometimes say 'Don't memorize this result;† it is easy to reconstruct it when you want it.' (Example: the equation of the tangent to some particular curve.) It is also certain that the initial process of memorizing is very much easier for symbols with meaning than for meaningless material.

In mathematics, what we *store* is a combination of conceptual structures with associated symbols, and the former would therefore seem to be important for the *retention* of the whole. The question is, which part of the symbols-and-concepts combination is easiest to 'catch hold of' when we are trying to *recall* material from this store into consciousness? And although there seems to be some evidence that it is by symbols that recall is most easily achieved, this view being consistent with the other functions of symbols, the situation is not entirely clear. Further research is needed.

(x) *Creative mental activity*

In one sense, since all new learning in mathematics by the method of concept-building consists of the formation by individuals of new ideas in their own minds, it is creative from their point of view. This is why, learnt in this way, mathematics is an exciting pursuit. But the description is used more particularly for the creation of ideas which no one else has had before – for opening up new paths, rather than retracing existing

* Or one of its equivalent forms.
† Meaning, of course, 'the symbols representing this result'.

ones, though the latter are new to the learner following them for the first time. The former is an unreliable process and may take years. Once the new insight is achieved, it can be communicated in the ways already discussed to all others whose schemas are far enough developed in the right direction to be able to assimilate it.

Ghiselin, in his classic work *The Creative Process*,[1] has collected together reports from originators in many fields – musicians, writers, scientists, mathematicians. From these, what emerges perhaps most clearly is that this process will not perform to command. The central part of the activity is both unconscious and involuntary.

There is, however, a fair degree of agreement that a necessary preliminary is a period of intense concentration on the problem. Following this, there is usually a period when the problem is laid aside, so far as the conscious mind is concerned, a period of relaxation, other mental or bodily activity or sleep. Apparently, during this period, unconscious mental activity concerned with the problem continues, for suddenly an insight relating to the problem – perhaps a complete solution – erupts into consciousness, at a time when no deliberate work on the problem is in progress. This insight is accompanied by a strong feeling of pleasure, and, interestingly, an urge to communicate.

What part do symbolic processes have in this creative activity?

Since the central stage, in which existing ideas suddenly fit together in a new way to produce an altogether new idea, is unconscious and involuntary, it is impossible to say whether, or to what extent, symbols play an essential or a contributory part here. In the preceding and the succeeding stages, however, their function is essential.

The first stage is that of intense, often prolonged, concentration on the problem, in which all the relevant ideas are brought together and considered from many aspects and in many different combinations and relations to each other. (Not all, for at this stage the insightful combination is not produced.) During this period of reflection, symbols play an essential part, for it is by their use that we achieve voluntary control over our thoughts. It may well be that it is at this stage that the contributory concepts are raised to a high enough degree of activity for the ensuing synthesis at the unconscious level.

When the insight does occur, it may well attach itself spontaneously to suitable symbols, for this seems to be closely associated with the process of making conscious. But this is likely to be incomplete, and the symbolization has to be continued deliberately, to make possible the communication and recording of the results of the creative process. This formulation and recording, with which is closely associated the process of making fully conscious, is often described as a painful struggle.

Unfortunately, too, not all the ideas which arrive in this way fulfil their early promise. After the insight must come verification. In science, this means the testing of the idea by experiment. In mathematics, it means logical analysis, testing for internal consistency and consistency with accepted knowledge. This is again a reflective process, for which symbols are essential. They may also, if chosen with care (which unfortunately is not always the case), contribute importantly to revealing the new structure.

*

The length of this chapter (it is the longest so far) is a measure of the importance of symbols in the learning and use of mathematics. There are two ways in which a relevant concept may be evoked: by encountering an example, which evokes it intuitively and involuntarily; and by the use of an associated symbol, which makes possible voluntary control, communication and the recording of knowledge. English and mathematics have both been described by Bruner as 'a calculus of thought', and it is their symbol-systems which make them so. Without an appropriate language, much of the potential of human intelligence remains unrealized.

CHAPTER 5

Different Kinds of Imagery

As long ago as the 1880s, Galton (1822–1911) found that people differed greatly in their mental imagery. Some, like himself, had strong visual imagery; others had none at all, and thought mainly in words. This is as true today as it was then, and there are also individuals who have available both, though often with a preference for one or the other modality. (It is not, however, always easy to decide what kind of images people use, or indeed whether they have any at all.) In this chapter we shall be considering the two kinds of symbol, visual and verbal, which are used in mathematics, both in mental imagery and for all the other purposes served by symbols.

Visual symbols and verbal symbols

First, these terms need clarifying, for as soon as words are written down they become things to be seen not heard. Nevertheless, words begin as auditory symbols, and their primary mode of communication is by word-of-mouth not word-on-paper. A reader usually turns written words into sub-vocal speech (though teachers of rapid reading point out that this is time-wasting). So by 'verbal' we shall mean both the spoken and the written word.

Visual symbols are clearly exemplified by diagrams of all kinds, particularly geometrical figures. But into which category should we put algebraic symbols like these?

$$\int_a^b \sin x \, dx$$
$$\{x : x^2 \geqslant O\}$$

Basically these are a verbal shorthand. They can be read aloud, or communicated without ever taking a visual form. The first is read as 'The integral from a to b of sine x dx' (or, '... with respect to x'); and the second as 'The set of all values of x such that x^2 is greater than or equal to zero.' The advantages of the algebraic notation are, first, those of any shorthand – a saving of time and trouble. But this brevity also adds greatly to its clarity and power, since the individual ideas for which they stand are evoked in a much shorter space of time, favouring apprehension of the structure as a whole. There may be less tendency to read them sub-vocally, and there are certain visual aspects which will be mentioned later. But as further discussion will show, algebraic symbols have much more in common with verbal symbols than they have with diagrams and geometrical figures, and for the present they will be classed with the former. A supporting argument is the way that verbal and algebraic symbols are mixed: for example, 'If p is a prime number, then $p \mid ab \Rightarrow p \mid a$ or $\mid b$.' ('If p is a prime number, then p divides ab implies that p divides a or p divides b.')

Both visual and verbal symbols are used in mathematics, together and apart. Thus we find diagrams with verbal explanations and, say, trigonometrical calculations; we find curves together with their equations; but we also find page after page of algebra with no kind of figure or diagram. Indeed, a recent and highly thought-of book on geometry also contains not a single figure! It looks as if verbal (including algebraic) symbols are indispensable, but visual symbols are not.

Even if they are not indispensable, however, there is no doubt that visual symbols are often very useful and may be a great deal more understandable than a verbal-algebraic representation of the same ideas. One sometimes also has the impression that the avoidance of diagrams is a demonstration, perhaps unconscious, that the writer needs no such props to his thinking – an academic 'Look, no hands!'

A reasonable working hypothesis is that the functions which these two kinds of symbol perform are different, perhaps complementary. If this is so, we want to know what these functions are, with a view to using and combining them to best advantage. For, let it be repeated, the part played in mathematics by symbols is crucial (see again the list on page 64). So any improvement in our knowledge of how to choose and use symbols, and devise new ones, would have great potential value.

Visual symbols seem to be more basic, at least in their primitive form of representations of actual objects. As Piaget has shown, even our perception of an object involves a kind of concept, though quite a low-order one. When we see any object from a particular viewpoint on a particular occasion, this experience evokes a memory of all our earlier experiences of seeing this object – not separately but as an abstraction of something common to this class of experiences. This is experienced as 'recognition', and we endow the object, in the present experience, with various other properties which derive not from the incoming sense-data but from the object-concept which is evoked. So a visual image, or a pictorial representation, of an object may fairly be described as a symbol, though the associated concept (that of the object) is of lower order than those used in mathematics.

By leaving out quite a lot of the visual properties of an object we can abstract at a higher level, while still representing the resulting concept visually. Maps, circuit diagrams and engineering drawings are all examples in which the most important properties of an object can be much better represented by visual than by verbal symbols.

For a mathematical example, consider this diagram, which represents a tall block of flats on level ground. For present purposes we are only interested in its height and shape.

Next we have represented a surveyor's observation of the angle of elevation of the top of the building, taken at a distance of 100 metres from the base. It is interesting here to note that the surveyor himself and the direction of his observation are both represented by spatial symbols (points and lines), while the measurements and the unknown height are represented by verbal-algebraic symbols.

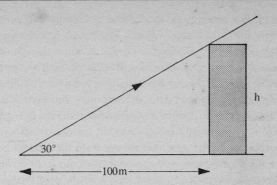

Already we need both, and as soon as the calculation begins, we go over completely to the latter.

$$h = 100 \tan 32°$$

Nevertheless, the diagram is of great help initially in representing the overall structure of the problem. It gives the context from which the particular calculation needed is abstracted.

Although they are more basic, visual images are much more difficult to communicate than auditory ones. For the latter, all we have to do is to turn our vocal thinking into speaking aloud. But to communicate our visual thoughts, we have to draw or paint or make a film. Nowadays, computer graphics make this easier, but the process is still much harder than speaking. This gives verbal communication a great advantage over visual. Moreover, the bringing into consciousness of an idea is closely linked with the use of an associated symbol. Since we hear our own speech at the same time as does the listener, the same ideas are evoked nearly simultaneously into the consciousness of both – provided, of course, that the symbols used have approximately the same meanings for both. So when speaking our thoughts to another, we are also communicating them to ourselves.

There is also some evidence that audible speech brings ideas into consciousness more clearly and fully than does sub-vocal speech. When working more difficult problems in arithmetic, children often drop back into whispering their thoughts; and in his voyage round the world in *Gypsy Moth IV*, Sir Francis Chichester found, when working the boat in difficult conditions and when very tired, that it helped to tell himself aloud what was to be done. This would explain the common experience that after simply stating a problem (academic or otherwise) aloud, even to a hearer who makes no contribution other than to listen, we sometimes find a solution.

When a discussion takes place, we get this subjective effect on both sides, together with the interaction of ideas which is the more conscious purpose of those taking part. The resulting progress of thought can be considerable. Because it is so easy to transmit our verbal symbols, and so much harder to transmit our visual symbols – we have built-in physical apparatus for the former but not the latter – the double advantage described above is attached, in the experience of most of us, much more strongly to verbal symbols.

Socialized thinking

It follows from this that our verbal thinking is likely to be more social-ized, since it is to a greater extent the end product not only of our individual thinking but of that of others, and of interaction between the two. To see things, literally, from someone else's point of view, we would have to go and stand where they were, or receive from them a drawing or a photograph, whereas they can talk about what they see without our making a move, and we can both hear the same sounds while standing in different places and looking in different directions. Vision is individual, hearing is collective, at the concrete as well as at the symbolic level. And it is interesting to notice that when we do wish to emphasize individual rather than collective aspects of a set of ideas, we talk about a 'point of view'. Even 'aspect' is a visual metaphor. So a contrast is beginning to emerge between the two kinds of symbols along the following lines.

> Visual: harder to communicate, more individual.
> Verbal: easier to communicate, more collective.

A human being is a social animal, and the advantages of communica-tion are so great that the predominance, noted earlier, of verbal thinking might well be explicable on these grounds alone. But the advantage of communicability is an accidental one (we have built-in loudspeakers but not built-in picture projectors) and not intrinsic to the nature of the symbols themselves. Indeed, it is sometimes said that 'one picture is worth a thousand words'. If this is so, then instead of writing this book (about 90,000 words), the author would have spent his time better in making ninety pictures. With modern techniques of re roduction, this would have presented no difficulty of publication. Moreover, the written word loses the particular advantages of simultaneity for speaker and hearer which the spoken word has, and the interaction between them. So, is writing books and reading them, rather than drawing them and looking at the pictures, simply a habit taken over from the habit of conversation and discussion? Or are there also intrinsic advantages in the verbal-algebraic kind of symbol?

Visual symbols in geometry

Geometry suggests itself as a profitable context in which to investigate this question, since this is one of the areas of mathematics in which diagrams seem to have particular importance. We must note at once that the symbols involved are more abstract than a visual representation of an object. Even a life-size colour photograph of an object shows only a single aspect, and, to the extent that it evokes the concept of the object-as-totally-experienced, it could be described as a symbol for the object. Other representations abstract further, usually showing shape rather than colour, texture, size. Another degree of abstraction is found in drawings which represent not a particular object but a class of objects. Even photographs may serve this purpose – one which advertises a new model of car is intended to persuade us to buy not that particular car but one of a particular class of cars. We attribute to it every property common to all members of that set – acceleration, speed, comfort, etc. – but no particular quality, such as engine number, colour. The photograph is just as much a symbol for a *variable*,* in the strict mathematical sense, as, say, the words 'Daimler Sovereign'.

A major difference between the two kinds of symbol, photograph and words, is that one looks like a typical object of the set which it represents, whereas the other does not sound like it. So this visual symbol, at any rate, has a closer link with the concept than has the corresponding verbal symbol. The same is true of geometrical symbols. This is a geometric symbol:

* See page 213.

This is the corresponding verbal symbol: a circle.

The resemblance of the geometric symbol to its concept has both advantages and disadvantages. An advantage is that it evokes well the properties of the concept. This is especially so when we represent visually several concepts together. The diagram then brings into awareness the relationships between these concepts far more clearly than does a verbal representation of the same concepts.

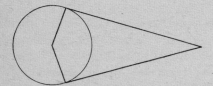

A circle, two tangents to it from a point outside the circle, and the radii through the points of contact of the tangents.

A disadvantage of the visual symbol is that it has to be drawn to be communicated – but pencil and paper, chalkboard and chalk, are easy enough to use. We also have to remember that it represents not a particular circle, tangent, etc., but variables – *a* circle, not *the* circle of given radius and centre which we see, etc. The words remind us explicitly of this. Since the diagram cannot but show a particular circle, etc., we have to remember to ignore its particular qualities and work with the general ones which it symbolizes. Because it is a stage more concrete, we must do some of the abstracting ourselves.

In the present example, however, these two minor disadvantages are quite outweighed by the conciseness and clarity of the visual symbols. Nevertheless we find that while most geometrical communications begin with a diagram, they very soon change over to verbal-algebraic symbols, together with additional geometrical ones such as $A\widehat{O}B$, $\perp$, $\parallel$. And the visual element is sometimes abolished altogether. In the study of vectors, directed line segments are replaced by ordered pairs, triplets or n-tuples of numbers;* and one of the directions in which geometry currently

* See Chapter 14, page 277.

seems to be moving is that of an algebraically manipulated axiom system. Why does not this, one of the most visual branches of mathematics in its early stages, remain so?

Visually presented arguments

The following examples suggest that we might, with advantage, stay in the visual mode more than we do at present. With a few simple conventions, the diagrams below convey all that the verbal statements do, more clearly and vividly.

The tangents to a circle from a point outside it are equal in length.

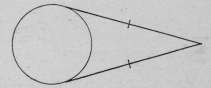

Note that the diagram also shows which parts of the tangents we mean, which in the verbal statement is left implicit and could only be made explicit by so many extra words that the meaning would then be less clear than before.

The exterior angle of a triangle is equal to the sum of the interior opposite angles.

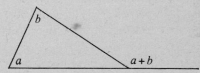

This is the usual statement. We should really say 'the size of the exterior angle', since an object and the size of an object are different ideas. This shows more clearly in the diagram, where the angles are represented by pairs of lines and their sizes by letters. And who would know which angles we meant by 'exterior' and 'interior opposite' without a diagram? Here the verbal statement is much inferior to the visual.

We can also show a theorem and its converse. The angle in a semi-circle is a right angle.

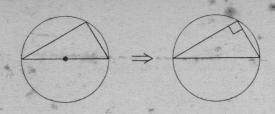

Here ⇒ means 'implies'. The left-hand figure shows the data, using the convention that a dot drawn approximately at the centre of a circle does in fact represent the centre. The right-hand figure represents the conclusion derived by this theorem from the data.

The converse of this theorem is also true. If a chord of a circle subtends a right angle at the circumference, that chord is a diameter.

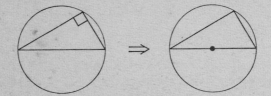

By using the sign ⟺ for a two-way implication, we can represent simultaneously both the theorem and its converse. The angle in a semi-circle is a right angle. Also, if a chord of a circle subtends a right angle at the circumference, that chord is a diameter.

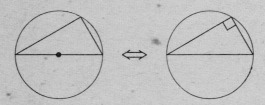

So far, the visual statements are much clearer and briefer. Difficulties begin to arise when we want to do two more things – give a logical proof and direct attention to particular parts of the diagram. The first of these often necessitates the second.

The above theorem is a particular case of the following. The (size of the) angle at the centre of a circle is twice the (size of the) angle at the circumference subtended by the same chord or arc.

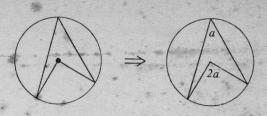

The proof of the earlier theorem consists in pointing out that we may

consider this straight line 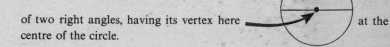 as an angle of the size

of two right angles, having its vertex here ⟶ at the centre of the circle.

The theorem given last tells us that this angle

is twice the size of this angle.

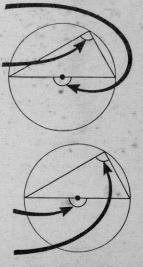

But the size of this angle is two right angles,

so the size of this angle is one right angle.

This is still clear, but much clumsier. In a face-to-face situation the same diagram would be used throughout, and the speaker would point to the parts of the diagram being referred to at the appropriate moments. The stumbling-block is the translation of an act of pointing into a diagram. Once we have drawn an arrow, we cannot erase it in a way which corresponds to the withdrawal of one's hand; we have to re-draw the diagram. And the arrows also clutter the diagram, because they are too like part of it. Different colours would help.

Another use of words has been to suggest new classifications to the reader: for example, that a straight line may be considered as a particular kind of angle. This can also be shown visually.

It takes more space, but is more vivid. There is a certain resemblance to a strip-cartoon, and if one has the resources and ability to translate this into computer graphics, the visual presentation can retain all its advantages. What would be the stages of such an animation? The following is one possibility. Note that the first figures represent the data.

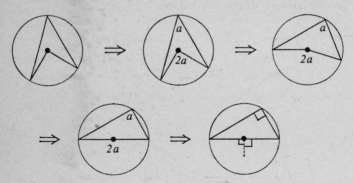

For comparison, here is a conventional proof of the same theorem.

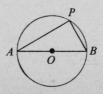

Data   *AOB* is a diameter of a circle, centre *O*.

P is any point on the circumference.

To prove $A\widehat{P}B$ = 1 rt ∠

Proof $A\widehat{O}B$ = 2 $A\widehat{P}B$ (∠ at centre = twice ∠ at circumference)

But $A\widehat{O}B$ = 2 rt ∠ s,   since *AOB* is a straight line.

∴ $A\widehat{P}B$ = 1 rt ∠

Q.E.D.

Here we use letters as a substitute for pointing. When the letters are found in the (verbal-algebraic) proof, we then have to find these letters in the diagram, and this tells us where to look. This is neater than the long arrows used on page 97 and saves re-drawing the diagram. Which is the easier to follow, the reader must judge for himself. This too could with great advantage be translated into computer graphics.

How does the 'purely visual' approach cope with more complex proofs? Space must limit us to one further example – a proof of this more general theorem already referred to.

Theorem

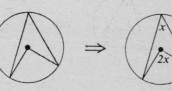

Proof

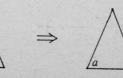

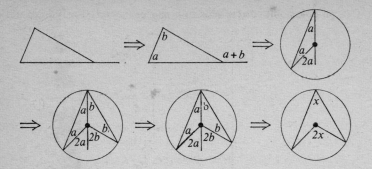

Is this clearer than a verbal-algebraic proof (for which, see any traditional school geometry text), or is it another case of 'look, no hands' – this time, no words? Since individuals differ in their preferences for visual or verbal-algebraic symbolism, there may be no general answer to this question. At present the latter system has achieved dominance. The chief purpose of the foregoing has been to question this *fait accompli* and examine the particular contribution of visual symbolism.

The two systems in conjunction

Historically, one of the happiest marriages of the two systems is that due to Descartes (1569–1650). Any point in the plane of the paper is specified by its distances from two (usually perpendicular) lines, that is, by two numbers, written as an ordered pair. These *coordinates*, as they are called, may be positive or negative.

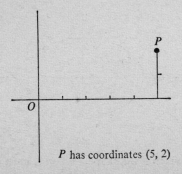

P has coordinates (5, 2)

A variable point corresponds to a pair of numerical variables; and a

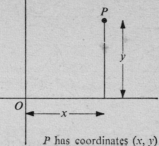

P has coordinates (x, y)

set of points with a given characteristic property, for example, that their distance from the origin is always equal to r, is represented by an equation satisfied by all the pairs of coordinates (x, y). By these means curves

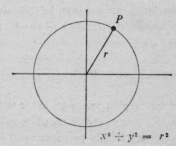

$$x^2 + y^2 = r^2$$

which are difficult to draw accurately can be represented algebraically: for example, an ellipse, which is the shape of a planet's orbit round the

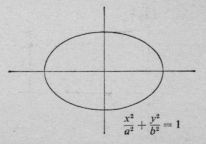

$$\frac{x^2}{a^2} + \frac{y^2}{b^2} = 1$$

sun; a parabola, which is the shape a reflector must be to give a parallel beam (as for a car headlight) or to concentrate distant rays to a point (as for a radio telescope).

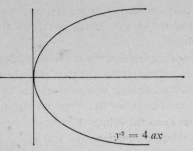

$y^2 = 4ax$

Both general and metrical properties can be dealt with in this way: general properties, by using general relations between variable coordinates, and metrical properties, by giving particular numerical values to these variables. What this algebraical treatment of geometry adds is great power of manipulation and accuracy far beyond what is available by accurate drawing to scale and measurement of the drawing. But we still need the drawing to show what the set of points looks like as a whole. It is, for example, not obvious from the equations that the curve represented by $y^2 = 4ax$ disappears into the distance in two directions, or that the curve represented by $\dfrac{x^2}{a^2} + \dfrac{y^2}{b^2} = 1$ joins itself again, or that a simple change of sign in the latter will give us something looking completely different.

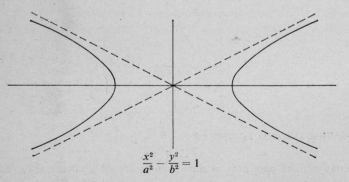

$$\frac{x^2}{a^2} - \frac{y^2}{b^2} = 1$$

That neither kind of representation is superior in all ways is suggested by the fact that we often use the method in reverse. Instead of starting with a known curve (all the above were known to Greek geometers about eighteen centuries before Descartes) and representing it algebraically, we may start with an algebraic concept, that of a function, and represent it graphically.

The idea of a mathematical function is one of great generality.* Broadly speaking, functions tell us how the objects in one set correspond to those in another: for example, how the distance travelled by an object may be found if we know the time; how the current through a given circuit may be determined if we know the voltage. Functions may be represented in a variety of ways, including equations and graphs.

For finding individual correspondences, an equation is very convenient. For example, if d metres is the distance travelled by a body in free fall under gravity (neglecting air resistance) and t seconds the time it has been falling, then $d = 4.9\ t^2$. So the distance fallen after one second is 4.9×1 metres, after two seconds it is 4.9×4 metres, and so on. By taking (t, d) as Cartesian coordinates, we can show graphically the function as a whole.

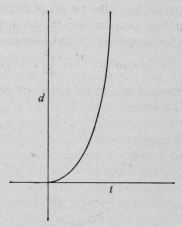

The two systems compared

Tentatively, we may now attempt a summary of the contrasting, and largely complementary, properties of the two kinds of symbol.

* It is discussed at greater length in Chapter 13.

Visual	*Verbal-algebraic*
Abstracts spatial properties, such as shape, position	Abstracts properties which are independent of spatial configuration, such as number
Harder to communicate	Easier to communicate
May represent more individual thinking	May represent more socialized thinking
Integrative, showing structure	Analytic, showing detail
Simultaneous	Sequential
Intuitive	Logical

The communicable, socialized properties of the verbal-algebraic system have doubtless contributed to its predominance over the visual system. Yet whenever we want to represent also the overall structure of some topic, argument or situation, visual symbolism returns, as in organization charts (from firms to football teams), flow diagrams and family trees. The value of visual symbolism is also shown by the way in which it super-imposes itself on the verbal-algebraic, in the form of spatial arrangement of written symbols. Auditory symbols are inevitably sequential in time. When written down, they are present simultaneously, the sequential arrangement being restored by scanning them in a conventionally agreed order. But this order may be departed from whenever we like. We may look quickly at the beginning and conclusion of an argument before ex-amining details. We may recapitulate whenever we wish, and this becomes necessary more often as the argument becomes more involved. In other words, a verbal-algebraic exposition, once written down, shows the over-all structure in addition to the logical-sequential implications within the structure, and it may be scanned in other ways besides the conventional left to right, top to bottom order.

Spatial symbolism finds its way into every detail of the verbal-algebraic system.

The position of a digit helps to show what number it represents.	2 7 3
	2 hundreds, 7 tens, 3 units.

Position shows which number is subtracted from which,	$9 - 5$
or divided by which.	$\dfrac{16}{4}$

Position shows correspondence
between two sets, as in
this proportion.

1	2	3	4	5
4	8	12	16	20

Its spatial arrangement
is an essential property
of a matrix.

$$\begin{pmatrix} a_1 & a_2 & a_3 & a_4 \\ b_1 & b_2 & b_3 & b_4 \\ c_1 & c_2 & c_3 & c_4 \end{pmatrix}$$

Many other examples could be given.

Before concluding this chapter, it will be interesting to return briefly to the individual differences in imagery noticed by Galton and mentioned at the beginning. If we are right in thinking that visual imagery is that most favourable to the integration of ideas, and if it is not accidental that when we first become aware of how ideas relate to each other, we refer to the experience as insight, not as in-hearing, then we might reasonably hypothesize that persons who have been noteworthy for their contributions to mathematical and scientific understanding will be found to use visual rather than auditory imagery.

We may as well begin the list with Galton himself, who tells us that his own visual imagery was clear but that he lacked verbal fluency. Einstein (1879–1955), in a letter to Hadamard, states that his preferred imagery is visual and motor, and that 'conventional words or other signs have to be sought for laboriously, only in a secondary stage'.[1] A famous non-mathematical example is that of Kekule (1829–96), whose conception of the ring structure of the benzene molecule came to him in a dream where he saw a snake taking hold of its own tail. And the Nobel Prize winner Bragg (1890–1971), in a television programme honouring his eightieth birthday, stated that his new ideas came to him for the first time in the form of visual images.

This list is a partial and selective one, and we lack the comprehensive information about other famous mathematicians which would support or refute the hypothesis. An interesting discussion, along more general lines, of personality traits of mathematicians and others is to be found in

Appendix 2 of Macfarlane Smith's *Spatial Ability*;[2] where much other material of interest in the present context may also be found.

Analysis, logical argument and socialized thinking are, rightly, much valued in mathematics, but we also need synthesis, insight and individual thinking. To some extent the former seem to be capable of being taught; the latter, at present, can only be sought. If we can discover more about the functions of the two kinds of symbol discussed in this chapter, and become more skilled in choosing and using them, this might well help us to develop and relate these two complementary aspects of our mathematical thinking.

Hemispheric specialization

Nine years after the first edition of this book was published, I attended a lecture by Glennon, subsequently published as a monograph entitled *Neuropsychology and the Instructional Psychology of Mathematics.*[3] Readers may judge for themselves the interest with which I heard, and subsequently read, the following passage. (To make even clearer the correspondences with the table on page 104, I have taken the liberty of interchanging the left- and right-hand columns in Glennon's table.)

In general, the left hemisphere processes verbal and analytic information. In general, the right hemisphere processes visuospatial and Gestalt (holistic) information. A summary of the findings from many research studies suggests the hemispheres perform these functions:

Right hemisphere functions	Left hemisphere functions
Visuospatial (including gestural communication)	Verbal
Analogical, intuitive	Logical
Synthetic	Analytic
Gestalt, holistic	Linear
Simultaneous and multiple processing	Sequential
Structural similarity	Conceptual similarity

Until this lecture, Glennon and I had not met, nor were we acquainted with each other's work.

CHAPTER 6

Interpersonal and Emotional Factors

This is primarily a book about learning mathematics with understanding, not about teaching it, though there are, of course, many implications for the latter. But most readers are likely to have the same attitude to the subject as they acquired at school, so an examination of these attitudes, and how they may have been acquired, is still relevant. For those with feelings of dislike, bafflement or despair towards mathematics, the aim of this chapter is to suggest that the fault was not theirs – indeed, that these responses may well have been the appropriate ones to the non-mathematics which they encountered. And those who remember their school mathematics with interest and pleasure will realize, if they did not before, how lucky they were. Chapters 1 and 2, in particular, have emphasized the particular dependence of the student of mathematics on good teaching, especially in the earlier stages, when foundation schemas, and also what may be long-lasting attitudes to the subject, are being formed.

Before contact with the students (whatever age they are), the teacher of mathematics has two important tasks: first, to make a conceptual analysis of the material; second, to plan carefully ways in which the necessary schemas can be developed, with particular attention to stages at which restructuring of the learner's schemas will be needed. Then, when in direct contact with students, the teacher is responsible for general direction or guidance of the work, for explanation and for correction of errors. The teacher also needs, to a varying extent, to create and maintain interest.

The before-contact tasks will usually be done by someone other than the face-to-face teacher. They are difficult and time-consuming, and the teacher who is involved in the day-to-day work of teaching is seldom in a position to undertake them. Whoever does – college lecturer or writer of mathematical texts – plays an essential part in the teaching process, but let us here for convenience restrict the term 'teacher' to the face-to-face teacher (or possibly correspondence-course tutor) who is in direct and continuing contact with the learner. In this chapter we shall be concerned with the personal interactions between teacher, in this sense, and learner, and the ways in which they may affect the learning based on understanding of mathematics.

What criterion?

Mathematics has much in common with the natural sciences and less in common with languages and subjects like history, English literature. It differs from all of these, however, in one important respect. In the natural sciences, the basic criterion for the validity of any statement or piece of work is experiment. Admittedly, not all the experiments will be done, or even witnessed, by the students. But in the main, if they are willing to accept in good faith that certain events result if certain conditions are set up, and particularly if they have some basic schemas based on their own experiments and observations, students of the natural sciences develop their knowledge in an interpersonal situation where the ultimate appeal is to facts and not to the authority of the teacher.

This is in marked contrast to some other subjects, for example, Latin, where the correctness of a piece of translation is decided on the authority of the teacher, or English, where again the final arbiter of the merits of an essay is the teacher (or examiner). In the former example, the teacher's opinion may be supported by the printed word, but this too is based on authority, not experiment. In the latter case, no appeal is available at all, except perhaps to another teacher – a second opinion, not an objective verification.

Where does mathematics stand in this? The question is important, because few people really like being told they are wrong, or otherwise diminished. But students are likely to accept this more readily if they can be given better evidence than 'because I say so', whether expressed thus or more politely. So what is (or should be) the final criterion for the validity of a mathematical piece of work – solution of an equation, proof of a theorem or answer to a problem in mechanics?

Certainly in pure mathematics, the ultimate appeal is not to experiment. (By what laboratory experiment can one prove that the square root of -1 is not a real number?) Nor is it, or rather nor should it be, to the teacher's authority. The final criterion of any piece of mathematics is consistency. This may be within a particular piece of mathematics – any solution to an equation must satisfy the equation in its original form, and if students offer an incorrect solution, this is how any averagely good teacher will tell them to check. Or it may be within the larger mathematical system of which it forms part. Whether this consistency exists is a matter for agreement between one mathematician and another, and between teacher and learner. The interesting, and rather surprising, thing is the high degree of agreement which can be achieved on such a basis. What is more, the criterion is implicitly accepted as binding by teachers and students alike. If a teacher makes a mistake when working on the blackboard and a member of the class points it out, the teacher has no alternative but to correct it. Teachers are subject to the same rules as pupils, and these are not the rules of an authoritarian hierarchy but of a shared structure of concepts. In mathematics perhaps more than any other subject the learning process depends on agreement, and this agreement rests on pure reason.

Insults to the intelligence

Students have no need to accept anything which is not agreeable to their own intelligence – ideally they have a duty not to. And it is by the exercise of the teacher's intelligence, not by prestige, eloquence or tyranny, that the students should be led to agree with their instructor. The teaching and learning of mathematics should thus be an interaction between intelligences, each respecting that of the other. Students respect the greater knowledge of the teacher, and expect their own understanding to be enlarged.

Suppose now that what they encounter is not intelligent or intelligible material at all, but a series of meaningless rules: for example, that they must, to solve an equation, 'get all the x's on one side and all the numbers on the other', and that the way to do this is to 'take them over to the other side and change the sign.' (See page 112.) Instructions of this kind may fairly be described as a series of insults to the intelligence, for they purport to be based on reason but (usually) are not.

The term 'insult' is used here both in the everyday sense and in the medical sense of something injurious to an organism. Trying to understand something involves assimilating it to one's schemas. To the extent that what is being communicated is not intelligible, the receiver is trying to expand or restructure personal schemas to assimilate meaninglessness. To do this would be equivalent to destruction of these schemas – the mental equivalent of bodily injury.

Viewed in this light, one can begin to see why some students acquire not just a lack of enthusiasm for mathematics but a positive revulsion. What is more, they are in these circumstances quite right in so doing, because one of their highest faculties, their developing intelligence, is being exposed to a harmful influence. That the teacher means no harm, but is only acting in ignorance, does not affect the situation at the receiving end. And it is, moreover, likely to be the more intelligent students whose minds boggle at the unorganized collection of rules without reasons which often constitutes the teaching of so-called mathematics. They are aware that they cannot find meaning in what is presented to them, but unaware that the fault is not theirs. Either the matter, in the form presented to them, is not meaningful, or else they have not been given certain preliminary ideas necessary to the understanding of the new ones.

Rules without reasons

This kind of teaching is as though someone learning to drive was told that whenever they wanted to come to rest they had to depress the clutch pedal as well as the brake, without ever having been told what was the function of the clutch pedal. 'Why?' they ask. 'If you don't, the engine will stop.' 'Why?' 'It just will.' The first reason is sound as far as it goes; but to answer the second 'Why?', two basic facts are needed. First, that an internal combustion engine will not, like an electric motor or a steam engine, start from rest under load. It has a minimum operational speed. Second, that to allow the engine to keep running independently of the gear box and road wheels, a gadget called a clutch is fitted which allows the engine to be connected to and disconnected from the gearbox at will.

'To divide by $\frac{2}{3}$, you multiply by $\frac{3}{2}$.' 'Why?' Readers are invited to search in their memories to find whether they have ever been given a good reason for this, or, alternatively, to seek an explanation from a school child of suitable age, to discover whether he or she has received any good reason.

The list of mathematical examples could be continued almost indefinitely, at both elementary and advanced levels. Some readers may remember learning to solve equations by some such method as the following, which is still in use: 'We use the rule that when we change the side we change the sign.'

To solve the equation we first get all the x's on one side by taking the x over and changing the sign.

$$6x - 3 = 7 + x$$

$$\therefore\ 6x - x - 3 = 7$$

Then we take -3 over to the other side and change the sign.

$$\therefore\ 6x - x = 7 + 3$$

Simplify both sides.

$$\therefore\ 5x = 10$$

Take the 5 across and change the sign.

$$\therefore\ x = 10 \div 5$$

$$\therefore\ x = 2$$

Answer: $x = 2$

If all that is wanted is to be able to solve equations of this kind quickly and efficiently, such a method is adequate. If, however, any importance is attached to understanding what one is doing, then it is not. And this understanding is not just a luxury which makes the task more pleasant: it is a necessity if one is to be able to adapt one's knowledge to new situations. The topology example given in Chapter 2 (page 44) was introduced to make just this point. In that example, the ideas which were necessary to convert the rule without reason into information which could be assimilated by the intelligence were few and simple. In the case of equations, the preliminary schema takes longer to build, so it will be deferred until Chapter 12.

Two kinds of authority

Whenever, and to the extent that, ideas prerequisite for understanding have not been made available to the learner, then whatever is communicated can only be in the form of assertions, and these will not provide nourishment for a growing intelligence. (The food metaphor is a close one. Genuine nourishment becomes part of the bodily self of the person who eats it; indigestible material is internalized, but not assimilated, and efforts to retain it indefinitely are contrary to our natural functions.) The acceptance of an assertion depends on the acceptance of the teacher's authority, and acting on it partakes more of the nature of obedience than of comprehension. In contrast, the assimilation of meaningful material depends on its acceptability to the intelligence of the student. Acting on it results from, and consolidates, enlargement of the student's schemas.

So far the word 'authority' has been used in what is probably its commonest connotation, that of a person to whom respect and obedience are due, as a result of status or function. Authority can also, however, result from superior knowledge, and this is, or should be, the kind of authority pertaining to a teacher as such. In schools (where we make our first, and some of us our last, attempts to learn mathematics), however, there is confusion and conflict between these two kinds of authority.

The former is closely related to the establishment and maintenance of discipline – of orderly behaviour and obedience to the teacher's instructions. This is the same kind of discipline, though of a milder kind (usually), than that imposed in the armed forces. But we talk also, though less commonly, of the disciplines of mathematics, chemistry, philosophy, etc. When a great scholar attracts disciples, they come as learners, and when they obey, it is willingly, because they want to learn.

School teachers have to exercise both kinds of authority, and promote both kinds of discipline. If they fail to control their young students, who do not attend school of their own free will, they have little chance of teaching them. Yet basically these two roles are not only different but in conflict. In other circumstances they are usually separated. At a meeting of a learned society, the former role is exercised by the chair, who calls the meeting to order, indicates whose turn it is to speak, and in general controls the conduct of the meeting; the other speakers – invited guests or participators in the audience – inform and discuss. It is improper for anyone to act contrary to the authority of the chair, but entirely proper for anyone to question and discuss the remarks of any of the other speakers, however eminent.

The combining of both these functions in one person may be necessary, but it is certainly unfortunate. In matters of orderly behaviour, performance of assigned tasks, choice of subject-matter, it is my view – which some will consider old-fashioned – that the students should accept the controlling role of the teacher, whereas the learning with understanding of the subject-matter thrives on questions and discussions among students and between students and teacher. Usually a reasonably satisfactory *modus vivendi* is reached, in which students learn how far teachers, in their first role, allow and even encourage them to express disagreement with them in their second role. Even so, artful pupils may use the second as a disguised form of opposition to the first, while teachers may subjectively experience a genuine request for explanation as a questioning of their (controlling) authority, and react inappropriately.

This role conflict matters particularly in mathematics for the reasons given earlier, that for this of all subjects, its learning and teaching need most to be based on reason and agreement. The situation will be aggravated whenever teachers are unable to give good reasons, because (perhaps through no fault of their own) they do not know them, and whenever (for lack of an adequate conceptual analysis) they have not developed the students' schemas in such a way that the material is experienced *by them* as reasonable. In these conditions learning based on understanding breaks down and is replaced (if at all) by learning based on respect and obedience.*

Benefits of discussion

So far we have centred our attention on the teacher–learner relationship. But discussion with fellow-students can also be an important contribution to learning. The mere act of communicating our ideas seems to help clarify them, for, in so doing, we have to attach them to words (or other symbols), which makes them more conscious. 'A problem clearly stated is half solved', and we have all found on occasion that in the process of formulating some problem, personal or academic, to a willing listener, we ourselves arrive at a solution. I met a teacher who uses an interesting technique when, in discussion, students make misstatements. A common response is to ask another student to explain to them where they are wrong. This teacher, however, asks students to explain to the rest of the class the reasons for their statement. The usual result is either that they discover their own error after a few sentences, or that the rest of the class learns something new.

* I wish to make clear my personal view (with which some may disagree) that, in their appropriate spheres, respect and obedience are necessary and desirable, since, without these, the conditions necessary for learning of the other sort could cease to exist. What I am trying to clarify is the distinction between two kinds of learning situation.

But there is more to a discussion than just thinking aloud. Another factor is the interrelating of our ideas with those of others – the expansion of our own schemas to enable us to assimilate their ideas, and the explanation of our ideas to them to enable them to assimilate our ideas to their schemas. Both are demanding, in different ways. The former requires flexibility and open-mindedness; the latter requires the ability to see just where the differences between one's own schema and that of the learner ('to see things from the other's point of view') lie, in order to know what explanation is necessary to bridge the gap. But if we can meet these demands, our own schemas will become enlarged thereby. More important still, they become more flexible: that is to say we, as total personalities, acquire habits and attitudes which favour further growth of our schemas.

Discussion also stimulates new ideas. One factor is simply the pooling of ideas, so that the ideas of each become available to all. Imagine a jigsaw puzzle in which the pieces are distributed among several persons, none able to see those of the other. Each might be able to complete part of the puzzle, or the pieces of each might be quite disconnected. But spread the pieces out on a table where everyone can see all the pieces, and they can all work together at fitting them together to form a meaningful whole.

The cross-fertilization of ideas is another benefit which comes from discussion. Listening to someone else (or reading what they have written) may spark off new ideas in us which were not communicated to us by the other, but which we would not have had without their communication. These may then, in turn, spark off new ideas in them, the result being a creative interaction which, at its best, can be exhilarating to all concerned.

Probably the best numbers for a creative discussion of this kind are small – two only, or at most three. Sometimes a new and fascinating idea is evoked, but before one can grasp hold of it the other person says something else, unwittingly distracting one's attention, and the fleeting glimpse is lost. A friend has suggested to me that there should be signals in discussion whereby either party may ask for silence if it is needed. Pencil and paper would also help the future retrieval of the idea and allow talk to resume. In this way a situation would be established unifying public discussion, private thought and written notes. This suggestion requires, and describes, good personal relationships between those taking part, and so points the way to another aspect of discussion.

Attitudes within groups

These benefits of discussion are also very dependent on friendly and fairly informal personal relationships between the members of the group. This does not mean quite informal. There must be certain agreed forms of behaviour, such as a willingness to take turns to speak, to listen, to consider the viewpoint of others. These are important parts of civilized discussion and are not too easily achieved. We do not much notice these forms of behaviour, because they facilitate the main task and do not intrude themselves on it.

If we do not like the fellow-members of our group, we are unlikely to be interested in sharing ideas, relating our own to theirs, or looking at things from their points of view. Rather the opposite – both we and the others will, according to temperament and circumstances, either try to make the group conform to our own ways of thinking or insulate ourselves from the pressures of the rest of the group.

This does not mean that the members have to agree in all their ideas or viewpoints; it means that they have to disagree in the right kind of way. That is, they have to agree that they will conduct their discussions on a rational basis and will neither make, nor react to, attacks on their statements or arguments as if these were attacks on themselves. And they have to agree on the final goal of any discussion – a step forward, by all, in the understanding of the subject.

The teacher as a group leader

An attitude such as that described above is a very mature one, which by no means all who take part in discussions achieve, whether children, adolescents or adults. And we also know, all too well, that people in groups can be much *less* creative, more destructive, sometimes even less human, than their members individually. Indeed, this seems to happen more easily than the creative interaction we have been discussing.

Which factors are at work is not yet fully known. Research on Freudian lines suggests that some of these are unconscious. Two of them, however, are clear – size and leadership. A large group degenerates into a mob more easily than a small one, and the part which each individual can take in discussion diminishes rapidly with size. My own experience is that quite small groups, numbering from two to five or six, are best, and although thirty to forty is a more usual number for a school class, there is currently a trend, particularly in primary schools, towards either individual work or work in small groups.

Where traditional class teaching is used with a fairly large class of students who are not there by their own choice, there is a situational pressure on teachers to take up an authoritarian attitude. If they do not establish and maintain order, they cannot fulfil their function as communicators of knowledge. Nevertheless, these two roles are basically in conflict, as indicated earlier, and the larger the group, the greater the conflict.

Ideally, a good teacher has to be both sergeant-major and conductor of an orchestra, able to alternate between these roles as required. To combine this with a knowledge of the subject is asking a great deal. I once watched a lesson in which the teacher achieved all three. Her control of the class was so well established and effortless that this role was not noticeable at all. In the course of the lesson a girl gave a wrong answer. The teacher wrote it on the blackboard; then by skilful questioning she led the class as a whole not only to find the right answer but to learn more from the wrong answer than they would have if the first answer had been correct. Moreover, the girl who gave it was not made to feel ashamed of, or embarrassed by, her mistake. It was also interesting to sense the intra-group feeling at the stage when about half the class had seen the point while half had not. Those who did understand showed, on their faces, the pleasure which accompanies a new insight; but they were also genuinely concerned to try to 'pull the rest of the class over the stile'. When everyone understood, there was a general relaxation of tension and feeling of satisfaction. The handling of her class by this teacher so impressed me that, at a subsequent meeting of teachers, I asked her to tell us how she did it. After a few minutes it was clear that she did not know, consciously. Her skilled group leadership functioned at the intuitive level and not yet at the reflective level.

Those who really understand mathematics are not common; those who can communicate it, less so; those who are also excellent group leaders, fewer still; while those who can also communicate this last ability are rare indeed.

Anxiety and higher mental activity

Another reason why the right kind of interpersonal relationship is so important in understanding mathematics is that anxiety itself may increase, subjectively, the difficulty of understanding. Given an exposition which, though not excellent, is more or less adequate, some pupils will be able to understand it, some not. If those who do not understand feel over-anxious at their failure, they will no doubt make greater efforts to comprehend. But this over-anxiety can be self-defeating, in that it can actually diminish the effectiveness of their efforts. The more anxious students become, the harder they try, but the worse they are able to understand; and so, the more anxious they become. Thus a vicious circle may be set in operation. Indeed, two: the short-term one just described and also a long-term one. Given several experiences of this kind, the situation itself, the mathematics lesson or lecture, becomes a learnt stimulus for anxiety; so the student begins each lesson already partially defeated. That this is not an exaggerated picture will be vouched for by many from their personal experience.

Here are some arguments in support of the belief that anxiety reduces – or may in certain conditions reduce – efficiency of mathematical thinking.

A principle known as the Yerkes–Dodson law has now, on the basis of experimental evidence, been fairly generally accepted by psychologists. This law states that the optimal degree of motivation for a given task decreases with the complexity of the task. In other words, for a simple task, the stronger the motivation the better the performance. But for a more complex task this is only so up to a point. Starting from zero motivation, which presumably produces zero performance, increasing the motivation improves the performance. But beyond a certain degree of motivation, further increase produces no further improvement of performance, but a deterioration. And the more complex the task, the lower the degree of motivation which gives the best performance.

Motivation is a fairly tricky thing to assess accurately, though performance is usually straightforward. This is because motivation is internal to the person concerned and not directly observable, while performance is observable and can be objectively assessed. To assess motivation experimentally, we have to set up conditions which we assume will have certain motivational effects on the subjects. For example, in one experiment rats were required to solve discrimination problems under water. They were confronted with two different doors, one of them locked, the other open and leading to air. The level of motivation was here varied by keeping them submerged for 0, 2, 4 and 8 seconds before they were allowed to start. Three different levels of difficulty of problems were used, and the results were in accordance with the Yerkes–Dodson law.

Understandably, there is less evidence of this kind available concerning human subjects. But let readers imagine themselves in a field when they discover that a bull is advancing menacingly upon them. The fiercer and closer the bull, the better their performance will be at running (a task of low complexity), jumping a ditch or climbing a gate. But suppose that the bull breaks through the hedge and readers seek safety in their car. Then, in the slightly more complex task of finding the right key and unlocking the car, they might well fumble. If the key were not in its usual pocket, they might take longer to remember that they had hidden it where others of the party could also find it if they returned first. Or suppose, by a stretch of the imagination, that they had to solve an easy problem to escape (as did the experimental rats), readers might well find that they took longer to do this than they would have under more relaxed conditions.

That the higher mental activities are the first to be adversely affected by situational anxiety has long been recognized by the army. Actions which have to be done under battle stress are taught as strongly formed habits, to be performed automatically, while those who have to plan the strategy of the battle and direct its tactics are kept out of the firing line. Many teachers, recognizing that examinations are a stress situation, similarly drill their students in well-practised routines.

My own experiments in this field have been based on the hypothesis that it is the reflective activity of intelligence (see Chapter 3) which is most easily inhibited by anxiety. One task used to test this hypothesis was a simple sorting task. Cards were prepared having one, two, three or four figures of the same kind on each. These figures could be squares, circles, crosses or triangles, and they were either red, green, yellow or blue, all figures on a card being alike and of the same colour. Four category cards were laid out: one red triangle, two green squares, three yellow crosses, four blue circles. The subject was given the remaining sixty cards and asked to sort them into piles in front of the category cards according to a single criterion. For example, a card having four green crosses would, if the subject had been told to sort by colour, be placed in pile two from the left. If sorting were by shape, it would be put in pile three; if by number of figures, pile four.

When the same criterion was used throughout, the subjects performed the task rapidly and efficiently. Moreover, their speed increased with practice. Subjects were then asked to sort the first card by colour, the second by shape, the third by size, the fourth by colour, and so on. This was no longer a routine task, but one involving reflective activity, albeit of a simple kind. (The reader may find it helpful to refer back to the diagram on page 54.) The subjects had to be aware of the category in use, this category being something internal to their own minds, not something external, and they had to switch this category to the next in series after each card had been sorted. The first is a receptor, the second an effector, activity of the reflective system.

The subjects were asked, as for the first task, to sort as fast and as accurately as they could. But under these conditions, far from improving with practice, they got steadily worse. Sometimes they broke down altogether – that is to say, they suffered a sort of 'mental blockage' during which time they could make no progress at all with the task. One subject, a university student of high intelligence, reported 'waves of panic, which I had to fight back'. The subjects were aware that they were being timed and that their errors would be noted, but were not otherwise subjected to external stress. It was quite striking how changing the task from a routine one (after a single reflective act at the beginning, to 'set up' the chosen sorting category) to one involving continual reflection was sufficient to produce conditions in the subject leading to moments of apparent mental paralysis.

It seems possible that the progressive effect might be due to the vicious circle described earlier. The worse the subjects performed, the harder they tried, and so the worse they performed, with consequent mounting anxiety. If this hypothesis were correct, then the interpolation of a simple routine task would interrupt the cumulative effect, and performance at the reflective activity would improve. This hypothesis was tested in a group experiment* with fifteen-year-old grammar-school boys, and it

* Details of the first experiment have been given in full, since it is an easy one for readers to repeat if they wish. The tasks for the group experiment were different, and for reasons of space are not given here.

was found that the progressive decline in performance was in fact removed.

Most of us can recall occasions when we have experienced a similar kind of momentary mental blockage. After important interviews, perhaps we feel that we could have given a better account of ourselves. Remembering that an interpolated routine task helps to reduce anxiety, I often, when interviewing candidates for university admission, begin with very straightforward questions and then interpolate more of these at intervals throughout the interview. Similarly a good teacher can, by initially asking questions that the student can answer, reduce anxiety and build up confidence, and thereby improve the performance; a bad teacher can reduce an averagely intelligent pupil to tongue-tied incompetence.

Here we are back to the interpersonal situation, and when considering the learning of mathematics it is difficult to keep away from it for long. Even adult students, learning independently from a text, cannot escape the historical effects of early teachers on their own self-confidence, or lack of confidence, in the mathematical-learning situation. When teaching elementary statistics to psychology students, the writer found that with many of these his first task was a remedial one, to convince them that they were indeed able to comprehend mathematics. I hope that readers who have unhappy memories of past attempts at learning mathematics will be willing to accept, as a working hypothesis, that the causes were other than their own lack of intelligence.

Initial causes of anxiety

In the last section it was suggested that anxiety, once present, could bring about a vicious circle of cause and effect in the mathematical-learning situation. On the principle that prevention is better than cure, we should now look for the causes which may introduce anxiety in the first instance.

One of these has already been suggested – an authoritarian teacher. This includes, of course, the strict disciplinarians of the old school. But in a milder sense, we must also remember that whenever the schemas necessary for comprehension are not present and currently available in the mind of the student, whatever learning takes place can only be based on an acceptance – a willing acceptance, perhaps, if the teacher is well-liked – of the teacher's authority. Learning of this kind is rote-learning not schematic-learning. Initially it may not be accompanied by anxiety, perhaps quite the opposite. Well-memorized multiplication tables, resulting in a column of neat red ticks, are rewarding to teacher and student alike. The problem here is that a bright and willing child can memorize so many of the processes of elementary mathematics so well that it is difficult to distinguish it from learning based on comprehension. Sooner or later, however, this must come to grief, for two reasons. The first is that as mathematics becomes more advanced and more complex, the number of different routines to be memorized imposes an impossible burden on the memory. Second, a routine only works for a limited range of problems and cannot be adapted by the learner to other problems, apparently different but based on the same mathematical ideas. Schematic learning is both more adaptable and reduces the burden on the memory.

Students of the kind described therefore inevitably reach a stage at which their apparent success deserts them. Try as they may, they can no longer 'get all their sums right'. The efforts they make are, of course, along the wrong lines – of trying to remember more and more rules and methods. Really they need to go back to the beginning and start again on new lines. Were this possible, the well-learnt routines would stand them in good stead (see page 82). But neither they nor their teacher know what is the matter, and even if they did there would probably not be time.

Here indeed is an anxiety-provoking situation, and there are now two vicious circles likely to be set in operation. The first is that described in the last section; the second is that the increasing efforts the student makes will inevitably use the only approach which he knows, memorizing. This produces a short-term effect, but no long-term retention. So further progress comes to a standstill, with anxiety and loss of self-esteem.

Adaptations to anxiety

Two important qualifications must now be made to the foregoing. The first is that the Yerkes–Dodson law refers to motivation in general, and we have so far been concentrating on motivation by anxiety. This is by no means the only, or the best, motivation. The second is that the optimal level of motivation for a given task will depend on individuals as well as the task. This is to some extent implicit in the earlier statement, that the optimal level decreases with the complexity of the task, for what is a complex task for one group of people may be a relatively straightforward one for another. The greater competence of the latter will help them in two ways: they will feel less anxious because they know they can cope; and they can use their anxiety constructively at work on the problem. A certain amount of anxiety can be a useful stimulus, and part of the background of education is to learn to use it as such. This we may call 'adaptation to anxiety'. Part of this adaptation results from having techniques for resolving anxiety-provoking situations – solving the problem or passing the examination in our present context. But another part is a personality variable. As such, it falls outside the scope of this book, but it is well worth noting that many of those who have contributed to knowledge have been not without their personal problems. We may speculate – but at present it is only a speculation – that in some way these people have found in their work an escape from anxiety, perhaps because it is an impersonal situation, perhaps also because it contains problems which they can solve.

Motivations for learning

So far our efforts have been directed towards trying to understand some of the factors which affect the learning and understanding of mathematics, on the assumption that we, or the students concerned, want to do so. But now, let us in all seriousness ask the question: why should anyone want to learn mathematics? It is indeed arguable that this question should have come right at the start of the inquiry, since without some kind of motivation there would be no reason to expect anyone to make the necessary effort. However, if you have bought this book, you probably have some kind of motivation. So let us now look at what this might be – or these, since several motivations may combine in a single activity.

'Motivated' is a description we apply to behaviour which is directed towards satisfaction of some need. If we say that a certain piece of behaviour seems motiveless to us, we mean that we do not know, and cannot even guess, what need is satisfied by means of it.

So questions about motives are usually, in disguise, questions about needs.

Some needs, such as food, warmth, sleep, are innate. Others, such as tobacco, soap, a television set, are learnt. Mathematics seems fairly obviously to be a learnt need; so how (if at all) do people learn to need mathematics?

One way in which new needs can be acquired is by learning that they lead to the satisfaction of already existing needs. In our present culture we soon learn that if we have money we can use it in many different ways to satisfy a wide variety of needs. Mathematics is also a valuable and general-purpose technique for satisfying other needs. It is widely known to be an essential tool for science, technology and commerce, and for entry to many professions. These are goals which motivate many adults to mathematics, but they are too remote to be applicable to the early years of school, when we first begin mathematics. In the classroom, shorter-term motivations are more likely to be effective: two of the most directly applicable here are the desire to please the teacher and the fear of displeasing her or him. Reward and punishment are widely used as methods for training both children and other young animals, and are older than schools themselves.

Both of these kinds of motivation are extrinsic to mathematics itself, however. Teachers can be pleased, or their displeasure avoided, by emitting the desired behaviour (verbal or written) with little or no understanding, so there is no guarantee that understanding has been achieved. Indeed, since understanding may take longer than parrot-learning, extrinsic motivation of either kind may favour the latter because it brings prompter results – quicker approval or quicker relief from anxiety, as the case may be. Of the two, motivation by anxiety is probably the more conducive to rote-learning, because, as we have already seen, it has an inhibitory effect on the reflective activity of intelligence.

Intrinsic motivations

But there are some people for whom mathematics is a pleasurable and worthwhile activity in itself, regardless of any other goals which it may also serve. These are the people whom I regard as true mathematicians, and if this view is accepted, then some seven-, ten- and twelve-year-olds merit the description as much as many sixth-form and adult students. Why people should enjoy learning and practising mathematics for its own sake is, however, far from obvious if we keep to our original hypothesis that any motivated behaviour satisfies some need.

Let us approach the problem indirectly, by way of other examples. Look at a child, out for a walk with parents, balancing along a low wall in preference to walking along the pavement. Or look at a dinghy sailor, sitting precariously out over the water in preference to the greater certainty and convenience of an outboard motor. Or look at a mountaineer, laboriously and hazardously climbing a mountain, which he could ascend quickly and safely by a funicular railway. Wall-walking, sailing, mountaineering are not basic needs, but neither do they apppear to be used as means to other goals, since in each of these examples there is a simpler and more direct means of attaining the end.

The apparent contradiction can be resolved if we hypothesize another very basic, very general need – a need to grow. The word 'grow' is used here to include not only physical growth but growth in skill, power, knowledge and any other physical, sensori-motor or mental organization which actually or potentially favours survival. Young children do not need to balance on walls, climb trees, jump off climbing frames, do forward rolls. But all of these serve, very directly, their growth needs: they develop the lungs, muscles and bodily control.

Mental growth is even more important for survival than physical growth, and activities which contribute to mental growth should therefore be enjoyed by children at least as much as physical activities. Mental growth, moreover, can continue long after physical growth has ceased, so the pleasures which come from the various ways of exercising one's intelligence should continue from childhood to old age. If it is agreed that genuine mathematics is simply a specialized form of intelligent activity, then we need no longer wonder why it can be enjoyable for its own sake.

The enjoyment we experience from activities, physical or mental, which serve our growth needs are experienced as intrinsic in the activity itself. Children don't like climbing because they know that it makes them strong and agile. Rather, they grow strong and agile because they like climbing. What is more, letting children climb trees is a better way to help them become strong and agile than making them do exercises. The rewards of doing something one enjoys are immediate and conducive to prolonging the activity itself; the more distant the goal, the greater the imaginative span required to relate present activities to it, the slower the apparent progress – in relation to the whole distance to be traversed – and, in general, the weaker the motivation.

For an adult, an excellent learning situation is one in which short-term and long-term motivations are fused, the short-term one being an enjoyment of the learning and doing of mathematics – an intrinsic motivation – and the long-term one being some personal, practical or academic goal to be achieved with the help of a knowledge of mathematics. But of the two, intrinsic motivation is probably the more important. Some things we learn because we know that they are useful. But the major strides which have been made, in mathematics as in the sciences, have resulted from the quest for knowledge for its own sake. Faraday is said to have replied to a woman who saw his demonstration of the deflection of a compass needle by a coil of wire through which an electric current was passing and asked what use that was: 'Madam, of what use is a new-born baby?' One characteristic of a baby is that it will grow, and another is that we cannot predict into what kind of adult. Even Faraday (1791–1867) could hardly have guessed at the long-term results of his discovery, whereby the connection between magnetism and electricity was established.

Similarly, a tendency towards growth is an intrinsic quality of the kind of mental organization which we call a schema. That we experience pleasure from any activities which are favourable to their growth is the most powerful incentive to learning, mathematics or any other subject. That the knowledge will afterwards be useful, or in what way, cannot be predicted at the time of learning, any more than when I buy a screwdriver I know exactly what jobs I am going to do with it. When they were studying calculus and algebraic geometry in college, the mathematicians of the American space research programme did not know that they would be using their knowledge to plot orbits for a lunar module.

How effective an intrinsic motivation for learning mathematics can be is something which many teachers do not yet appreciate. On a number of occasions, teachers finding that children actually enjoy mathematics when it is intelligently taught and learnt have reported this to me with a mixture of surprise and pleasure; but also of doubt, as if something must be wrong with an approach to mathematics which children enjoyed. But until this intrinsic motivation is better comprehended and put to work, mathematics will remain for many a subject to be endured, not enjoyed, and to be dropped as soon as the necessary exam results have been achieved.

PART B

Introductory Note

As originally planned, this book concentrated entirely on the mental activities involved in learning mathematics with understanding, and would have finished at the end of the present Part A. It was then felt that a book about the understanding of mathematics, with so little mathematics in it, seemed incomplete; so it was decided to add a second part, applying the ideas of the first part to some of the basic topics of mathematics.

At this stage, several difficulties arose. Part A was written for the intelligent general reader (an old friend of many authors) with no prior knowledge of psychology; it did not matter whether little or much mathematics was known, since the mathematical examples were illustrations of the psychological principles, useful if the reader was already in possession of the relevant mathematics, but not essential. Part B, however, could not be written on this basis. To learn mathematics *ab initio* requires the provision of collections of examples from which new concepts can be formed, and also further examples, for practice and consolidation of concepts at one level of abstraction before they are used as examples from which further concepts of higher order can be formed. Practice is also required to become familiar with the symbols used. Books which provide these are textbooks, of which there are plenty already in existence. They take time to work through, and their aim is not only comprehension of the ideas but mastery of the techniques. It did not seem that this approach was what was wanted here, from which it follows that Part B could not be written on the assumption of no previous mathematical knowledge at all.

The present aim is different: to try to illustrate the development of mathematical schemas, starting with a conceptual analysis of some of the most basic ideas. For the earlier part of this, however, only the most elementary knowledge of mathematics is necessary. It may be described as 'elementary mathematics from an adult viewpoint', examining critically some of the ideas and activities which we already have, but which for the most part we take for granted. This can take us from a stage of simplicity (in which we do things without thinking about why they work) to one of doubt (in which we become aware of much that we do not understand) before we emerge again into a new state of simplicity resulting from comprehension. If at any time readers feel that they are stuck in the second of these stages, my suggestion would be that they put the book aside for a while and return to it later. Since most mathematical concepts are of high order, the thinking involved is very concentrated, and one may soon suffer from an information overload. This is one consequence of not doing examples, which keep one for a while at the same conceptual level. It is also an indication to read more slowly. It may also be the case that one understands something while reading it, but for lack of practice does not have it firmly enough established to use it as a starting point for further ideas. Some back-tracking may therefore be necessary.

Anyone with 'O' level mathematics should have no great difficulty with any of the chapters which follow. For these, it will be largely a matter of analysing and perhaps restructuring ideas with which they are already familiar. For those who have forgotten most of the mathematics they once knew, the earlier chapters should still be straightforward, but the later ones will probably be difficult. So please take from this section whatever you can read with interest and pleasure. And so persistent are our attitudes and habits that it may not be inappropriate to add a reminder that there is no examination to be taken at the end! So there is no need to assimilate any more than is found to be worthwhile for its own sake.

CHAPTER 7

Beginnings

Most people, if asked what are the ideas with which mathematics begins, would reply 'numbers' or possibly 'counting'. We shall therefore begin our conceptual analysis with these two closely connected ideas. Before beginning, however, it is perhaps worth warning the reader that the ideas which will be introduced are elementary in the sense of basic, but not necessarily in the sense of easy. It is sometimes harder to explain something apparently simple (how does a wheel work?) than something more complex.

Number and counting

Number and counting are by no means inseparable. It is possible to have a rudimentary idea of a number without being able to count, and Piaget has shown that children can count in a restricted sense (which will be described later) without really having the concept of number. But if by counting we mean something like 'finding the number of apples in a bowl', then it is clear that counting in its everyday meaning is a way of finding a certain property of a collection of objects, which we call its number.* This implies that number and counting are ideas which belong closely together, and that, of the two, number is the more basic.

Sets

But we have already implied an even more basic idea. Before we can count them, we must know which objects we are to include and which to

---

* When the names of numbers are used simply as a convenient list of words to use as names, for example, P.C.49, M & B 693, the word 'number' is not being used with the mathematical meaning which we are here trying to analyse.

exclude – that is, which objects belong to the collection of objects we are for the moment concerned with. Suppose now that, in a roomful of people, we are asked to count the number of good-looking girls. A difficulty presents itself: whom do we include? And supposing that we are able to decide according to our own judgement, will others necessarily agree? We are already in danger of losing, at the outset, one of the essential bases of our mathematics – agreement on rational grounds.

So, for mathematical purposes, we must reject requests like the above and agree to confine our attention to *well-defined* collections of objects: that is, collections such that given any object, we can say whether it is in the collection or not. Such a collection we shall call a *set*; and the objects which are in it we shall call its *elements*. (The words 'belong to' are also used, in this context, interchangeably with 'are in'.)

Characteristic property of a set

How can we 'well-define' our collection of objects? The method which has been used in the examples so far is to use some kind of adjective or adjectival phrase – a brief description. If the object fits this description, for example, of being both an apple (not a pear or a peach) and in the bowl (not on my plate or on the tree), then it belongs to the set. If it does not fit this description, then it does not belong to the set. A property (in this example, a double property) which defines membership of a set is called the *characteristic property* of that set.

Another way of defining a set is by listing its elements. For example, we could define a set as consisting of the moon, my left ear and the Nelson monument. Having thus defined a set, we could go on to say that it has a characteristic property – that the names of its elements appear on the list. A characteristic property of this sort does not usually lead to new ideas. The more interesting sets are those from whose elements we can abstract a new concept. But listing is nevertheless a perfectly good way of defining a set, as is any other method which satisfies the criterion given above. It also illustrates the two directions in which we can go, in our thinking, between sets and their characteristic properties. We can decide on some criterion first, and collect in our set all the objects which satisfy the criterion: for example, the set of European capital cities or the set of oak trees in the British Isles. Or we can form some collection of objects, in any way we choose provided that it gives us a well-defined collection, and then try to find what (if anything) they have in common. This method is fruitful in arriving at new concepts, and is the one we shall use in developing the concept of number.

There is no restriction on the kind of objects from which we form a set. They can be material objects (for example, the set of coins in my purse) or abstractions (for example, the seven cardinal virtues). We can even have a set with no objects in it at all, such as the set of pigs with wings. This is called 'the empty set' or 'the null set'. This set is well-defined: given any object at all, we can say whether or not it belongs to this set. (It does not.) We can also have sets whose elements are themselves sets. For example, 'Manchester United' is the name of a football team, a set whose elements are footballers. This set is an element of another set, whose elements are the teams in the Football Association. Examples of this kind are many, both in everyday life and in mathematics.

It is often useful to agree in advance which objects are under consideration for inclusion in, or exclusion from, a given set. Suppose, for example, that the teacher of a class of biology students is trying to get them to define for themselves the idea of a fish, and asks rather loosely 'Tell me something which is not a fish.' One student replies 'Mount Everest', another 'a tiger', a third replies 'a whale'. Why do we feel intuitively that the first answer is a silly one, the second not very good and the last answer a good one? Because it was certainly implied that the discussion was about animals, so the 'Mount Everest' answer, though true, is irrelevant. No one would consider 'tiger' as a possibility for inclusion in the set of fishes, so this example contributes nothing to the discussion either. But a whale might well be thought by some to be a fish, so this answer recognizes that the question really meant 'Tell me an aquatic animal which is not a fish.'

The questioner should, no doubt, have said so. Mathematically speaking, we would say: 'They should have defined their *universe of discourse*.' * This is the name given to the set of objects under consideration in any particular discussion or discourse, and once this has been defined, any other sets referred to must contain only elements which belong to the universe of discourse.

Another idea we shall need is that of a *sub-set*. Given, say, the set of

* Also called 'universal set' and sometimes 'universe'.

stamps in my desk drawer, I may divide these up into the sets of one-penny stamps, two-penny stamps, etc. These are all sub-sets of the first-named set. The sentence at the end of the last paragraph may now be restated: 'All sets referred to must be sub-sets of the universe of discourse.'

The importance of the idea of a set and of the related ideas which will shortly be developed is that they form a bridge between the everyday function of intelligence and mathematical thinking. From one side, the concept of a set is simply a recognition of something we do all the time when we classify the things we encounter. 'What's this?' means 'To what class, or set, of objects does this belong?' But once made explicit, the idea of a set, together with ideas that derive closely from it, will be found among the most helpful of all in clarifying the elementary, and many of the advanced, ideas in mathematics.

What do we mean by 'three'?

A reasonable rejoinder to the above question might be: why do we want to know? I can recognize three cups of coffee or three cows when I see them; what more do I need? Before continuing, readers are entitled to be convinced that what they will find, if they read on, is not a long and complicated answer to an unnecessary question.

A first answer is that there is good scientific precedent for questioning the obvious. Our present knowledge would be the poorer if no one had questioned why an object appears coloured or why the leg of a dead frog twitches under certain conditions. People did, it is true, get on quite well without asking these questions. But 'well is the enemy of better', and we are all better off today because of the inquiring minds of Newton and Volta (1745–1827).

A second answer is that we can, and do, use the concept 'three' effectively at the intuitive level without ever analysing it. Most of us can, and do, manage pretty well in everyday life without being (even in a small way) mathematicians. We use simple mathematical techniques without understanding them, just as many people use cars and television sets without understanding them.

Quite a small degree of understanding, however, greatly increases the effectiveness with which all of these can be used, and if the reader wishes to understand some of the beginnings of mathematics then this is hardly possible without an understanding of the concept of number, at the reflective as well as at the intuitive level. Negative numbers, fractions, irrational numbers, are all generalizations of ideas derived from the counting numbers 1, 2, 3 ... (called by mathematicians 'the natural numbers'). And the whole of elementary algebra starts from general statements about numbers. Number (by which in this chapter we still mean natural number) is a concept whose examples are particular numbers such as 1, 2, 3 ..., so let us consider what we mean by, say, '3'. Any other smallish number would do equally well.

A good way to understand an idea at the reflective level is to imagine how we would convey it to someone else. In this case, let us also assume that the person has not available concepts of the same or higher order which would enable us to use a definition successfully. (See pages 23–6.) Just as we would have to convey the concept *red*, in these circumstances, by pointing to various red objects, so in this case we would have to find an assortment of objects which exemplified the concept 3. In so doing we would find that *all the objects we chose were themselves sets*: three apples, three fingers held up, three pencils, three chairs. *Three* is the characteristic property of a certain collection of sets, of which (ourselves already having the concept) we can choose a sufficient variety to enable our student to form the concept.

Matching sets

By similar means we could also convey the concepts 4, 5, 6 ... To convey concepts such as 17, 48, in this way might also be possible, though it would certainly take longer. But another problem would then arise. Given that our students now know the meaning of the adjectives 1, 2, 3 ..., 17 ..., 48 ... how can they decide for themselves, given some new set, what adjective to attach: in other words, what is its number? The difficulty increases rapidly with the size of the sets.

They cannot yet count, of course, and in all probability this is what most of us would immediately set out to teach them. But if we proceed slowly, we shall arrive at counting nearly as soon, and shall on the way acquire some other ideas which will be helpful in the future.

Using again our 'red' example as a starting point, suppose someone wanted a simple way of deciding whether a new object encountered also had this property. All they would have to do would be to carry in their pocket any convenient red object and compare the object with it. If, and only if,* the colours matched, they would assign the new objects to the 'red' set. By also carrying in their pocket a green object, a blue object, a yellow object, etc., they could find the colour of any other (single-coloured) object they wanted by similarly testing it for match against the standard objects.

If this experiment were carried out in practice, it would soon be found that these sets having red, blue, green, yellow . . . for characteristic properties were not as well-defined as might be wished (even assuming for universe of discourse the set of all visible, single-coloured, objects). Matching objects for colour is not precise, and what some might consider as a rather greenish blue, others would describe as a bluish green.

Matching sets for number can, however, be done with complete precision, and we do it frequently in everyday life. Given a set of saucers on a tray, we match this with a set of cups, and very likely with a set of teaspoons, all these sets being of the same number. If this number is fairly large, we may not bother to find out what it is. We simply ensure that to each saucer there corresponds a cup, and vice versa. There must be a cup for each saucer and a saucer for each cup, with none left over of either set. Similarly for the set of spoons. A match of this kind is called a *one-to-one correspondence*, and whether any two given sets match in this way or do not is unlikely to be a matter for disagreement. It is also clear that if we have a one-to-one correspondence between the set of saucers and the set of cups, and between the set of saucers and the set of spoons, we shall also have a one-to-one correspondence between the set of cups and the set of spoons. A diagram makes this clear.

Given this match between a set of cups and a set of saucers

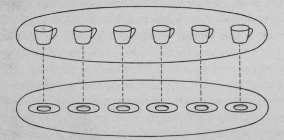

---

* This may conveniently be abbreviated to 'iff'.

and this match between a set of saucers and a set of spoons

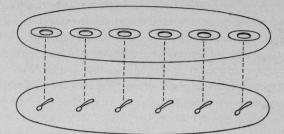

we also have a match between the set of cups and the set of spoons

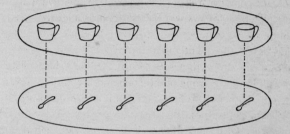

This property of one-to-one correspondence is called the *transitive property*. Using '↔' as an abbreviation for 'is in one-to-one correspondence with', and calling the sets of cups, saucers and teaspoons respectively C, S and T, we express the result briefly by saying that if $C \leftrightarrow S$ and $S \leftrightarrow T$, then also $C \leftrightarrow T$.

It follows from this property that, in the given example, we can use either the set of cups, or the set of saucers, or the set of teaspoons for comparison with a new set, say, a set of packets of biscuits. If the new set matches one of the existing sets, it matches them all. For if, say, $B \leftrightarrow C$, then, since $C \leftrightarrow S \leftrightarrow T$, we know that $B \leftrightarrow S$ and $B \leftrightarrow T$.

This result is true for any set of matching sets; so, for purposes of comparison, any one of a set of matching sets can represent this set (of matching sets) as a whole. And given any new set, it is easy to decide by the above matching procedure whether it belongs to a given set of matching sets or not. Our sets are now well-defined.

An interesting way of ensuring that two sets match, without knowing the number of either, was used in the days before general literacy and numeracy. Suppose that a master wished to ensure that the flock of sheep which arrived home from market was the same in number as when it departed. Notches were cut in a stick called a tally stick, each notch corresponding to a sheep. The tally stick was then split lengthwise, each half being the 'split and image' of the other. The master kept one half, the herdsman the other, by which means each 'kept tally' of the number of sheep.

This method uses the transitive property too. If S and S' stand for the sets of sheep which leave the market and arrive home, and T for the set of notches cut on the tally before it is split, then $S \leftrightarrow T$ and $T \leftrightarrow S'$ ensures that $S \leftrightarrow S'$ as desired. And since originally $T \leftrightarrow T_1 \leftrightarrow T_2$ (where T_1 and T_2 represent the sets of notches on the split tallies), if either T_1 or T_2 is altered *en route* there will be a mis-match at the end.

In symbols, provided that no sheep are lost or gained,

$$S \leftrightarrow T \begin{array}{c} \nearrow \ T_1 \ \searrow \\ \updownarrow \\ \searrow \ T_2 \ \nearrow \end{array} S'$$

Standard sets and ordering

The particular importance of the transitive property for our present purpose has also been mentioned: that in a class of matching sets, any one will do for testing a new set for inclusion in, or exclusion from, this class.

In the 'red' example we suggested that our students might conveniently carry around in their pockets an object of each colour with which to compare any new objects encountered. These would be their standard objects for colours. To find the number of any new set they would need to have with them a set of each number (a set of seven, a set of three, a set of nineteen . . .) with which to compare the new set. These standard sets for numbers they would choose on the basis of convenience and portability, so parts of the body might suggest themselves. For a set of two, they might choose eyes; for a set of five, the fingers of a hand; for a set of four, the limbs; for a set of one, the nose. Calling these sets 'eyes', 'hand', 'limbs', 'nose', would lead to these becoming names for the numbers which we call 'two', 'five', 'four', 'one'.

In this form, the standard sets would certainly be portable, but still somewhat inconvenient, for several reasons. First, their arrangement is haphazard. To find the number of a new set students would have to try it for match against whichever standard set they considered likeliest. If they were wrong, they would have to pick another. There is as yet no organization in their trials.

The obvious way to introduce this would be to try out the standard sets in a regular way, starting with that containing fewest objects, then the smallest of those remaining, and so on. That is, to arrange them in order. Since they cannot rearrange their anatomy, they would do this mentally, by putting the number-names in order: nose, eyes, limbs, hand.

Another deficiency now becomes obvious: that a standard set is missing between eyes and limbs. 'Clover' might suggest itself as an easily remembered set of three. So the ordered standard sets would now be: nose, eyes, clover, limbs, hand.

Counting

Just when the last step from the above to counting was taken we do not know; nor do we know whether it came in a flash of insight to some solitary genius or whether the invention was repeated at different times and places over the world. But both for its brilliant simplicity and effectiveness, and for its contribution to civilization, we must rank it equal with that of the wheel.

This step is to use the *number-names themselves*, in order, as standard sets. (Reminder: The curly brackets below denote a set.)

Number in our invented notation	*Standard set* containing that number of words
nose	{nose}
eyes	{nose, eyes}
clover	{nose, eyes, clover}
limbs	{nose, eyes, clover, limbs}
hand	{nose, eyes, clover, limbs, hand}

Finding the number of a given set of objects is still done in the same way as before, by putting the elements of a standard set in one-to-one correspondence with those of the given set. This is a particularly convenient arrangement, done by touching, pointing to, looking at or just thinking of the objects to be counted while reciting the number-words in order.

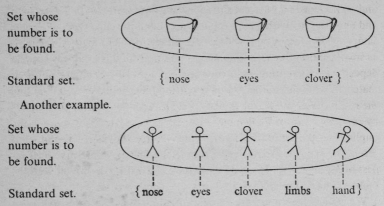

Set whose
number is to
be found.

Standard set.   { nose   eyes   clover }

Another example.

Set whose
number is to
be found.

Standard set.   { nose   eyes   clover   limbs   hand }

We no longer need to know in advance which standard set to try. Instead of a trial-and-error process, trying first too large a standard set, then too small, until we get it right, we simply pair the number-words with the elements of the given set until the latter are all used. *The last word we say is the number of the set of words we have used*, selected automatically.

This is what we are doing every time we count. Unfamiliar names for the first five numbers have been used temporarily, in an attempt to present a very familiar activity as if it were new to us.* We shall now revert to the familiar words and numerals. Once the standard sets have been ordered, we are very close to counting. But this does not diminish the importance of the last step, for while the numbers of quite small sets could be fairly easily found by trial-and-error matching, it is only by the technique of counting that we can find, quickly and accurately, the numbers of large sets; and they do not need to be very large before the superiority of counting asserts itself.

Counting and arithmetic

As an example of how dependent other mathematics is on counting,

* For this idea I am indebted to Miss Kate Sowden, Whitelands College of Education.

consider a simple statement like '7 + 5 = 12'. There is, as usual in mathematics, more in this than meets the eye, but let us leave aside for the moment the process of abstraction which leads to such a statement and embody it in a physical situation. 'We have a tray with 7 cups on it and another tray with 5 cups on it. If we put all the cups together on the same tray, how many cups will there be altogether?' The method whereby, as small children, we learnt to answer questions of this kind was entirely dependent on counting: first, counting out a set of 7 counters, cubes or matchsticks, then another set of 5, putting them together and counting the result. Later, we learnt to add (perhaps with the help of our fingers) by *counting on* from '7', five more words: '8, 9, 10, 11, 12'. 'Counting on' implies that the set of five is now being counted together with the set of seven. We now need some device, such as five fingers, to preserve the fiveness of the second set of words, since the last word we say will now be the number, not of the second set, but of the combined set.

We cannot here undertake a conceptual analysis of arithmetic along the same lines as that which has been done for counting, for to be as thorough at all stages would turn this into a many-volume treatise. The reader may convince himself by a few more examples of the dependence of arithmetic on counting, and he may care to analyse for himself some of the other activities of elementary arithmetic in terms of the concepts of sets, one-to-one correspondence and counting; for example, subtraction (is this always the same as 'take-away'?) and multiplication. In the next chapter we shall discuss another topic closely associated with counting, that of numeration – the systematic naming of numbers. But before closing this chapter let us look briefly at the relationship between counting and number as they are met by children today.

Children counting, and their concepts of numbers

These days children learn very early to recite in order the names of the numbers up to five or ten, usually before they go to school. They can pair these words with the objects in some set without many errors. So they can, while very young, often give correctly the numbers of various small sets. This is indistinguishable from counting, except for one small detail – they may not necessarily have the concepts of the numbers themselves!

Piaget[1] has shown this by a variety of ingenious experiments, one of which is summarized as follows. Two matching sets of six eggs in six eggcups were put before a boy. He was asked whether there were as many eggs as eggcups or not. He said 'Yes, there are.' Next, the eggs were taken out of the cups and bunched together, the eggcups being left where they were. When then asked whether there were the right amount of eggs to go into the cups, one in each and none over, a typical five-year-old boy among Piaget's subjects said 'No, there are not enough eggs.' Asked to count both eggs and cups, he did so correctly, saying that there were six of each. Again asked whether there were as many eggs as cups, he still said that there were not. He had not yet grasped two essential properties of any number, that it is a common property of two matching sets and that the match (and so the common property) is not affected by changes in position of the subjects.

It is interesting to compare this with the behaviour of a 3½-year-old boy who put some carriages of his toy train from the box on to the rails, saying 'One for me, one for you and one for mummy.' This child, though unable to count, was matching the two sets mentally, regardless of the position of the objects (his mother being out of the room). So which of these two children was closer to the concept of a number?

Collecting objects into sets on the basis of a common property (all the red ones together, all the blue ones . . .; or, all the pictures of fruit, all the motor cars, all the people . . .) is one pre-mathematical activity; ordering (by various criteria) is another; comparing two sets to see whether they match is another, on the borderline of mathematics.

Ideally, it would be preferable for children to have plenty of experience of all of these before learning to count. But counting is so much a part of the world around them that children learn to recite the number-names not long after they learn to talk, whether we like it or not. Does it matter?

Children copy from adults and other children many words and phrases whose meanings they acquire gradually afterwards. In my view it does not matter that children learn to 'count' before they really have the concepts of the numbers whose names they are reciting, provided that the latter concepts are at some time learnt and attached. The danger is that the transition to written work may be made on the basis of 'pseudo-counting' alone, without essential contributory concepts. In this case, the young learner's mathematics starts on the shakiest of foundations.

Given suitable opportunities, however, counting can contribute to the formation of number concepts in two ways. First, as has been shown on page 73, naming can help in the process of forming new concepts. Counting and recounting the same objects after several re-arrangements could lead to the suspicion that there was some property which persisted through these changes. Also, counting is both a matching process itself (though rather a specialized one) and an excellent way of testing two sets for match. 'Have we a cup for everyone?' The answer is quickly reached by counting cups and persons, using, let it be noted, the transitive property of one-to-one correspondence mentioned earlier.

The best way is probably to give children plenty of experiences leading to all the relevant concepts being gradually formed alongside each other and gradually connected into a schema. Strict hierarchic development does not seem to be essential – but plenty of concept-building, experimental and manipulative activity before pencil-and-paper work, certainly is.

Summary

In this chapter mainly everyday ideas have been taken, examined more closely and formulated in ways which bring them into mathematics as foundation concepts. The following are the chief ideas which we shall need to remember.

Set: a well-defined collection of objects, which may be physical objects or ideas.

Element: one of the objects in a set.

Characteristic property of a set: a property such that a given object belongs to the set iff it has that property.

One-to-one correspondence (also called in this chapter, for short, matching): two sets are in one-to-one correspondence iff to each element of one set there corresponds one and only one element of the other set.

A number: the characteristic property of a set of matching sets.

Standard set: any convenient representative of a set of matching sets used for comparison with a given set to find whether it belongs to that set of sets.

Counting: a device for finding the number of any given set by using the names of numbers, in order, as standard sets.

CHAPTER 8

The Naming of Numbers

In Chapter 7 we found that the names of numbers, when themselves collected into sets, could be used to give a quick and easy way of finding the number of any given set of objects. This is, or should be, a surprising use for number-names. It is as if the names of the colours could in some way be used to find out the colour of some object we are looking at, or a list of the breeds of dog to find out the breed of any given dog. This method, we found, would work equally well whether we used the familiar names 'one, two, three . . .', or new ones chosen by ourselves – 'nose, eyes, clover . . .' Does it, then, not matter how we name the numbers? In this chapter we shall see that it does, and that the Hindu-Arabic system which we use today contributes greatly to our powers of calculation.

Numeration

We shall use the term 'numeral' for any name of a number. Thus (as stated in Chapter 4) 5, V, cinq, 101, are different numerals for the same number, the last being a binary numeral. (To describe words spelt out, like 'five', as numerals is a slight extension of familiar usage, but since '5' and 'five' are both pronounced alike, and since both are symbols for the same concept, it seems more consistent and convenient to call both numerals than to call '5' a numeral and 'five' a number-name.) 'Numeration' means the naming of numbers, so the question is now whether one system of numeration is better than another and, if so, what makes it so.

The answer to the first part of the question soon becomes apparent if we try to add in Roman numerals, and even more so if we try to multiply. Compare these additions:

$$\begin{array}{ccc} \text{XXIV} & \text{and} & 24 \\ +\ \text{XXXIX} & & +\ 39 \\ \hline \end{array}$$

or these multiplications:

$$\begin{array}{ccc} \text{XXIV} & \text{and} & 24 \\ \times \quad \text{VII} & & \times \quad 7 \\ \hline \end{array}$$

or these long multiplications:

$$\begin{array}{ccc} \text{XXIV} & \text{and} & 24 \\ \times \quad \text{XXXIX} & & \times \quad 39 \\ \hline \end{array}$$

The Roman numerals do not even tell us at a glance what sort of size a number is. Though XLIX is one less than L, it doesn't look it! No wonder that even simple multiplication was a task for the professional mathematician, and that calculations had to be done with aids such as calculi (pebbles).

A major fault of the Roman system is that it fails to make use of the fact that if we add a set of 2 and a set of 3, we get a set of 5, whether the elements of these sets are single objects or themselves sets of objects or even sets of sets. Thus 2 matches and 3 matches together make 5 matches. Also 2 boxes of 40 and 3 boxes of 40 make 5 boxes of 40. Also, 2 packets of a dozen boxes of 40 and 3 packets of a dozen boxes of 40 make 5 packets of a dozen boxes of 40. So the one abstraction $2 + 3 = 5$ serves for all these operations. In the Hindu-Arabic system, this result is used in the form 2 sets of 10 together with 3 sets of 10 make 5 sets of 10; and similarly, 2 sets of 10 sets of 10 together with 3 sets of 10 sets of 10 make 5 sets of 10 sets of 10, and so on. It also says the foregoing much more clearly and concisely: $2 + 3 = 5$, $20 + 30 = 50$, $200 + 300 = 500$, etc. If you find the second set of statements easier to follow than the first, this will of course be partly due to familiarity, but it also demonstrates the value of conciseness in a notation for the easy handling of what is seen, by expansion into verbal sentences, to be quite a lot of information.

Whenever we deal with large numbers of objects, we tend to collect these into sets, and then to collect these sets into sets of sets, and then to collect these sets of sets into sets of sets of sets, and so on. For example, platoon, company, battalion, etc.; letter, word, phrase, sentence, para-graph, chapter, book, library. But a method may be widely used in one context, while its transfer to another remains long unthought of. The Romans organized their army along the same lines as those described, but not their way of writing numbers. They wrote II + III = V, but XX + XXX = L and CC + CCC = D – three results to remember instead of one.

In the Hindu-Arabic system, we learn as children to add all pairs of numbers of the form $n_1 + n_2$, where n_1 and n_2 stand for any of the numbers 1, 2, 3 . . . 9. Since any first number can be added to any second number, this entails memorizing 81 results, which reduces to 45 by virtue of the fact that $7 + 5 = 5 + 7$, etc. If this only enabled us to add pairs of numbers up to 99, a total of 4950 pairs, this would be a saving of 4905 facts to be remembered. But the Hindu-Arabic system enables us to add all possible pairs of numbers, as large as we like, simply by memorizing 45 facts of addition, together with a few simple rules of procedure, so the saving becomes incalculable. It is like the difference between walking and flying: journeys become easy by the latter means which could hardly be made by the former. And it is my belief that a major reason why Roman mathematics 'never got off the ground' was their lack of a good arithmetical notation.

To find out how the Hindu-Arabic system achieves this tremendous saving of labour we need to go back to first principles and examine with adult understanding some of the simple routines for adding and multiplying which we learnt as children. We shall find that the simplicity of the methods conceals an unexpected complexity of ideas, which may help us to appreciate the power which is conferred by the notation and associated techniques and in part where this power comes from. To limit the task we shall assume, without analysis, the idea of place value: for example, that 365 means 3 hundreds and 6 tens and 5 units.

Adding

In everyday speech we use 'add' for many different acts of combining: for example, 'Add a beaten-up egg.' Here, we need to distinguish between a way of combining two sets, which we shall call *uniting*, and a way of combining two numbers, which we shall call *adding*. So adding two numbers, say, 5 and 7, corresponds to

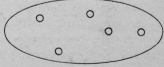

taking any set whose
number is 5

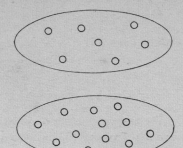

and any set whose number is 7

uniting these as one set,
and finding the number of
this new set. The resulting *set* is called the *union* of the original two sets;
the resulting *number* is called the *sum* of the original two numbers.*

We need a more concise way of writing all this. Let S_1 and S_2 stand for
any two (disjoint) sets and $S_1 \cup S_2$ for their union. Then, in general, if
$n(S)$ means 'the number of the set S', etc.,

$$n(S_1) + n(S_2) = n(S_1 \cup S_2)$$

This shows clearly the connection between uniting two sets and adding
their numbers. If the notation is unfamiliar to readers, it will be worth their
familiarizing themselves with it by constructing a few examples, such as:

$n(S_1) = 5$

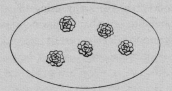

A set of 'Ena Harkness' roses.

$n(S_2) = 7$

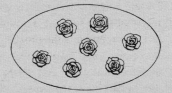

A set of 'Charles Mallerin' roses.

* It is here understood that the two sets have no element belonging to both: that is, that the
sets are *disjoint*. We can form the union of any two sets, disjoint or overlapping (for example,
the set of children and the set of females), but, in the latter case, some will be counted twice, so
what we say here about their numbers is only true for disjoint sets.

$n(S_1 \cup S_2) = 12$

A set of red roses.

The idea which needs consolidating, for future use, is that uniting the two sets of roses and adding their numbers involve three levels of abstraction: physically putting the two bunches of roses together, which is an act in the outside world; doing the same in thought, which is a mental act with primary concepts; and adding the numbers, which is a mental operation on secondary concepts. And the last is no less a mental, abstract, operation if we write it down; the symbols reduce the difficulty of working at this level of abstraction by helping us to control our thoughts.

The above notation will help us to remember that a number is a property of a set and not of its elements. When we refer to 'a set of blue cups', the adjective 'blue' describes the cups. But when we refer to 'a set of six cups', the adjective six describes the set, not the cups. This inaccuracy of everyday language does not matter at all for everyday purposes, such as buying cups – the meaning is clear. But if we are now going to build a conceptual hierarchy using these as basic ideas, we need to be more careful to say what we mean. So we should put the adjective 'six' next to the noun which it describes, and talk about 'a six-set of cups'.

Notice that when 'blue' describes the cups it is an adjective. But when, at the next level of abstraction, it refers to something which these cups have in common, 'blue' becomes a noun, the name of a colour. Similarly when 'six' describes a set, it is an adjective. (All the sets pictured on pages 138–9 are six-sets.) When it is the name of a number – the common property of these sets – it is a noun.

Just as 5 represents the property common to all sets which match the standard set {'one', 'two', 'three', 'four', 'five'}, so $5 + 7 = 12$ represents what is common to all acts of uniting sets like the above, whatever may be the particular sets involved. Since the result depends only on the numbers of the sets concerned, we can work with whatever sets (having these numbers) we find convenient: for example, fingers, matchsticks, Unifix cubes or o's on paper, as here. Having arrived at the result, it can be recorded, memorized and used as a short cut to give us the result for any union of two sets * with these numbers, without having to repeat the performance of actually uniting the sets and counting.

As beginners, we learn that any five-set united with any seven-set make a twelve-set. Abstracting, we say 'five and seven make twelve' or $5 + 7 = 12$. (The latter distinguishes more clearly between the two levels of abstraction.) It is by working with sets of physical objects that we first develop these and similar concepts. But consider now, say, $37 + 45$. To find the answer to this, do we have to take

a set whose number is 37,

a set whose number is 45,

unite these, and count
the elements of the new set?

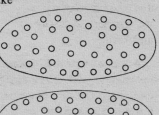

Clearly this method soon becomes unwieldy. By the Hindu-Arabic notation it is also made unnecessary, because as soon as we have written the number of the first set as 37, we have mentally sorted it into 3 ten-sets and 7 singles.

Similarly for the second set,
whose number is 45.

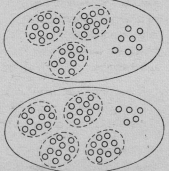

* Reminder: we are talking about disjoint (non-overlapping) sets.

After uniting two sets which have been arranged in this way, we only have to count small numbers of ten-sets and small numbers of singles, remembering that whenever we have ten singles we think of them as 1-ten-set.

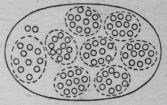

The union of the two given sets contains 7 ten-sets and 12 singles, which we rearrange as 8 ten-sets and 2 singles.

At the numerical level, the notation does this organizing for us. In any numeral, the right-hand digit * is the number of singles, the next reading from right to left † is the number of tens, the next is the number of hundreds, and so on.

So by writing the numerals in appropriate columns we add the units, tens, etc. This corresponds to counting together all the singles, all the tens, etc., in the union of the two sets. Adding seven and five mentally, we think 'twelve'. But the notation 12, together with a rule which allows only one digit per column (the 'carrying rule') automatically rearranges this as 1 ten and 2 singles.

$$
\begin{array}{cc}
3 & 7 \\
4 & 5 \\
\hline
 & 7 \text{ twelve} \\
1 & 2 \\
\hline
8 & 2
\end{array}
$$

This 'carrying' method works equally well from tens to hundreds (tens of tens). This is yet another application of the principle that the number of a set does not depend on what are its elements.

---

* One-symbol numerals, '1', '2', '3' ... '9', '0'.
† Which is the way Arabic writing is read.

Compare the foregoing addition with the same in Roman notation, which gives no help whatever of this kind. XXXVII represents the smaller number of the two, yet it uses twice as many

XXXVII
XLV
———

symbols. Three Xs represent three tens, which is reasonable though cumbersome. But for four tens, on the next line, a single L is used representing five of them, preceded by a single X meaning one ten less. Yet XV means ten *more* than V and XVII means two more than XV. There is little consistency, and the places of the symbols in the upper numeral and the lower bear no relation to each other.

Multiplication

The advantages of the Hindu-Arabic notation are even greater when it comes to multiplication. Let us start by considering what, say, $6 \times 3 = 18$ means* in terms of sets of objects.

If we start with a six-set

and combine 3 of these

the result can be rearranged as 1 ten-set and 8 singles.

* It is best to read this here as '6 multiplied by 3' for consistency with long multiplication, which follows shortly. The other reading, '6 times 3', is discussed briefly on page 159.

Clearly it makes no difference to this result if we put ten dots in each small o. All the singles become ten-sets and all the ten-sets become hundred-sets.

So

hundreds	tens	singles
		6
	×	3
	1	8

becomes

hundreds	tens	singles
	6	0
×		3
1	8	0

that is, $60 \times 3 = 180$.

Just as we can add numbers by the same method whether they represent singles, tens, hundreds, etc., so also we can multiply large numbers by a single digit at a time, carrying where necessary to the next column from right to left. Example: 586×3.

This condensed notation, while convenient, obscures what we are really doing. Most children learn it too soon and are encouraged to omit the 'carried'

$$\begin{array}{ccc} 5 & 8 & 6 \\ & & 3 \\ \hline 1 & 7_2 & 5_1 \ 8 \end{array}$$

figures as soon as possible. This favours mechanical speed at the expense of mathematical comprehension. We are really working three separate multiplications:

$$\begin{array}{r} 5 \ 8 \ 6 \\ \times \quad\ 3 \\ \hline \end{array}$$

$$
\begin{array}{rcl}
6 \times 3 & = & 1\ 8 \\
80 \times 3 & = & 2\ 4\ 0 \\
500 \times 3 & = & 1\ 5\ 0\ 0 \\
\hline
\end{array}
$$

Adding, $586 \times 3 \ = \ 1\ 7\ 5\ 8$

The distributive property

The last line of the foregoing calculation depends on the truth of the following:

$$(6 + 80 + 500) \times 3 = (6 \times 3) + (80 \times 3) + (500 \times 3)$$

Here, parentheses indicate which operations are to be done first – in each case, that within the parentheses. So the left-hand side instructs us to calculate $6 + 80 + 500$: result 586. Then, to calculate 586×3. The right-hand side instructs us to calculate 6×3, 80×3, 500×3, and then to add these three results. The equality sign asserts that the numbers obtained by these two methods are the same.

How are we to be sure of this, since the calculation on the right-hand side is the only one we know how to do? We know our $\times 3$ table up to 9×3 and we know that it is true whether we use it on singles, tens or hundreds. But we do not know our $\times 3$ table up to 586×3.

Any particular case can be checked by addition.

586×3
is the number
of the set we get
by uniting
3 sets, each of
number 586.

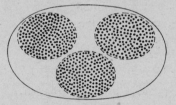

(Imagine 586 dots in each of the smaller sets.)

We can add these
numbers by the method
already established –

	5	8	6
	5	8	6
	5	8	6
1	7	5	8

though there are some more hidden assumptions, which we will look at later.

So the method gives the right answer in this case. But can we be sure that it will always do so, whatever the numbers? Given that a method works for as many individual cases as we care to examine, it is still a different matter to demonstrate that it must always work, of necessity. This we shall now do, but the method can be more easily seen by starting with some particular cases.

This figure embodies,
in sets of o's, the
multiplication 3 × 4.
(Start with a three-set,
and unite 4 of these.)
Rearrange it like this,
where each column
represents a three-set, and
the set of four columns
represents the result of
uniting four such sets. This
figure similarly embodies
the multiplication 2 × 4.
Uniting these two sets
corresponds to adding
the number of rows.
Without calculating, it is
clear that the resulting
figure embodies the
multiplication (3 + 2) × 4.
Hence (3 × 4) + (2 × 4)
= (3 + 2) × 4

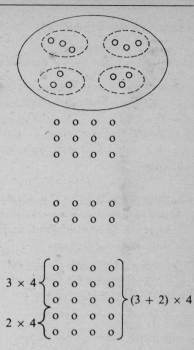

Since no calculation is involved, it is clear that this method does
not depend on which particular numbers we use. Nor do we even need to
know what they are. Let n, a, b, stand for any numbers whatever.

This figure embodies
the multiplications
    a × n

    and

    b × n

(a × n) + (b × n)
= (a + b) × n

←n columns→

$$
(a + b) \times n
$$

This property of natural numbers, when written the other way about, thus:

$$n \times (a + b) = (n \times a) + (n \times b)$$

is often stated in words as: 'multiplication is distributive over addition'. It is therefore called for short the *distributive property*. It is partly because of this property that we can calculate an indefinitely large number of products like 586×3 by knowing our $\times 3$ table only up to 9×3, and do likewise with other multipliers. So its importance is incalculable.

The two properties of addition

Are there any other properties which we take for granted, but on which all our well-established methods of calculation are also dependent? One of these is used whenever we add numbers greater than 10. We work the example on the right by calculating mentally:

$$\begin{array}{cc} 2 & 3 \\ 6 & 4 \\ \hline 8 & 7 \end{array}$$

$$3 + 4 = 7 \quad \text{and} \quad 2 + 6 \ = 8$$
$$\text{meaning} \quad 20 + 60 = 80$$

There are two hidden assumptions here. The method is only valid if

$$23 + 64 \qquad = (20 + 60) + (3 + 4)$$
$$\text{that is, if} \quad (20 + 3) + (60 + 4) = (20 + 60) + (3 + 4)$$

using parentheses as usual to show which operations are done first. This requires (i) that it does not matter which pairs of numbers we add; (ii) that the result is unaffected by changing the order of the numbers. These properties of the operation of addition on natural numbers we have long taken for granted. Formally, if a, b, c, are any numbers,

$$\text{(i)} \qquad (a + b) + c = a + (b + c)$$
$$\text{(ii)} \qquad a + b = b + a$$

In words, the first states that the result is the same whichever two numbers we associate first, and the second states that the result is the same if we commute (interchange) the numbers to be added. So these two properties may be stated briefly in words: *addition of natural numbers is associative and commutative.* When embodied in sets, these properties become intuitively obvious.

The first says that if we are uniting three sets the number of the final result will be the same whichever two we unite first. The second says that the result of uniting two sets is unaffected by order.

Multiplication is associative and commutative

It is natural now to ask whether multiplication of natural numbers has similar properties, and, if so, whether these are as useful.

That multiplication is commutative we have long taken for granted. Embodied in sets, however, the property can be seen as non-trivial.

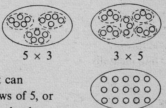

5 × 3   3 × 5

The set on the right can
be regarded as 3 rows of 5, or
5 columns of 3, and clearly
matches both of the first two sets, which therefore match one another.

The commutative property of multiplication is also implied in the alternative ways of reading 5 × 3 as '5 multiplied by 3', which corresponds to the top left-hand set of sets shown above, or as '5 times 3', often read as five threes, which corresponds to the top right-hand set of sets. One of its values has already been mentioned – it halves, approximately, the number of results of the form $n_1 \times n_2$ which need to be memorized. Another emerges from an analysis of long multiplication.

This is a method for
multiplying by 37 without
having to learn our 37
times table.

```
            4  1  2
               3  7
         _____
         2  8  8  4
      1  2  3  6
      _____
      1  5  2  4  4
```

It uses the distributive property in two ways. When multiplying 412 first by 7 and then by 3, each of these calculations depends on the distributive property, as discussed on page 158. Also, the extension of the method from multiplying by 3 to multiplication by 37 assumes that

$$(412 \times 37) = (412 \times 7) + (412 \times 30)$$

But how did we calculate 412×30 without knowing our 30 times table? We multiplied by 3 and then by 10, doing the latter by moving the result one column to the left. This assumes that

meaning
$$412 \times 30$$
$$412 \times (3 \times 10) = (412 \times 3) \times 10$$

that is, that multiplication is associative – the result is the same whichever two numbers we multiply first. Let us verify this in an example with smaller numbers, say, 3, 4, 5.

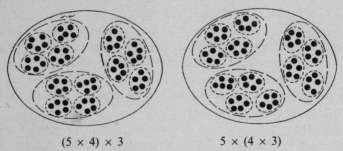

$$(5 \times 4) \times 3 \qquad 5 \times (4 \times 3)$$
(Each small set loop contains 5 dots.)

The only difference between these is that, for the calculation on the left, we remove first the dotted loops surrounding the 4 five-sets of dots and then the dashed loops, whereas for the calculation on the right we take off first the dashed loops surrounding the 3 four-sets. Clearly the numbers of the final sets, whose elements are the dots themselves, will be the same.

The five properties of the natural number system

By the natural number *system* we mean the set of natural numbers {0,* 1,

---

* Zero was a latecomer to the set of natural numbers.

2, 3 . . .} together with the two operations + and ×. We have found that the familiar methods for adding and multiplying depend on five properties, which may usefully be summarized here.

In words		In symbols
		If n, a, b, c, stand for any numbers, then:
Addition is	commutative	$a + b \qquad = b + a$
	and associative	$a + (b + c) = (a + b) + c$
Multiplication is	commutative	$a \times b \qquad = b \times a$
	and associative	$(a \times b) \times c = a \times (b \times c)$
Multiplication is distributive		$n \times (a + b) = (n \times a) + (n \times b)$
over addition		

Most of us take these properties for granted without appreciating their importance. They enable us to extend our ability to add and multiply from pairs of numbers less than 10, like $2 + 5$, 3×4, to sums and products of numbers of any size, like $24372 + 738946$, 192×205932. Though, happily, learning to do lengthy calculations of this sort with mechanical speed and accuracy is no longer regarded as an essential part of school mathematics, the machines which do them for us make use of the same properties. These, in turn, rest mainly on two foundations.

(i) The number of a set does not depend on what the elements are. In particular, the elements may themselves be sets.

(ii) The number of a set does not depend on how the elements are arranged, which is to say it does not depend on the order in which we count them.

Our ability to make good use of the properties of the natural number system has resulted from a notation which embodies these properties, and which leads to simple and rapid methods of calculation which exploit them to the full. Since our present commerce, industry and technology would be impossible without efficient methods of calculation, it follows that the material side of modern civilization has been made possible, to an important degree, by the Hindu-Arabic notation.

CHAPTER 9

Two More Key Ideas

So far, by 'numbers' we have meant natural numbers only. In Chapters 10 and 11 we shall expand our idea of number to take in four more number systems: fractional numbers, integers, rational numbers and real numbers. Two key ideas for the understanding of these are the ideas of *equivalence* and of a *mathematical model*. The importance of these ideas is not, however, confined to number systems. Like the idea of a set, they are fundamental.

Equivalence

This is one of the ideas which help to form a bridge between the everyday functioning of intelligence and mathematics, and it will be useful to start with everyday examples before defining it mathematically.

The word 'equivalent' suggests the meaning 'worth the same', that is, the same for a certain purpose or in a particular way. Given any set of objects, we can often sort this set further into sub-sets which are alike in some way. For example, {coins in my pocket} can be sorted into sub-sets of coins having the same value; {pots of paint in a certain shop} can be sorted into sub-sets of pots of paint of the same colour; {novels in the local library} can be sorted into sub-sets of novels by the same author. A method of sorting will be incomplete if there are any objects in the parent set which do not belong to one of the sub-sets, and ambiguous if any object can be assigned to more than one sub-set. So we say that every object in the present set must belong to one, and only one, sub-set. A set of sub-sets which satisfies this requirement is called a *partition* of the parent set.

This sorting of the elements of our parent set into sub-sets can be done in two ways. We can start with some characteristic properties, and form our sub-sets according to this. For example:

Sets	Characteristic properties of sub-sets
{coins in my pocket}	1p, 2p, 5p, 10p, 20p, 50p
{pots of paint}	red, blue, green, yellow ...
{novels in the local library}	H. G. Wells, C. S. Lewis, Neville Shute ...

Notice that the characteristic properties themselves usually belong together – they form a set which itself has an easily seen characteristic property. Thus in the first example, each characteristic property is a monetary value; in the second, each is a colour; in the third, each is an author. This need not be so. If we are standing on the pavement in London and are in a hurry to get to the station, then we may divide {passing objects} simply into the sub-sets {taxis} and {everything else}. But the examples for which it is so are usually the more interesting, since they form the basis of new ideas.

Alternatively, we can start with a particular matching procedure, and sort our set by putting all objects which match in this way into the same sub-set. For example, naturalists might sort {butterflies captured in a certain country} by matching their specimens by wing colour and pattern. Each sub-set of butterflies would be regarded as a different species and given a different name. This method is frequently used when encountering new objects.

A matching procedure of this kind is called, if exact, *an equivalence relation*. The necessary exactness can be achieved in the realm of mathematical ideas, but not so easily in the physical world. Suppose, for example, that we are sorting strips of wood by matching them for length. We might decide that strips A and B were of the same length if they differed by only 5 mm; and similarly for B and C, C and D, etc. But it would still be possible for planks A and J to differ in length by as much as 45 mm. So the match 'is approximately the same length as' is not transitive.* On the other hand the match between two sets 'is in one-to-one correspondence with' (see page 138) is exact, and is therefore an equivalence relation. Reverting to the planks, if we measure the lengths of the planks to the nearest whole centimetre, and match the planks not physically but according to these measures, it can be seen that the transitive property is satisfied, and so we now have another equivalence relation.

* This term was explained on page 139.

In addition to the transitive property, equivalence relations have two further properties which, for our present purposes, need not be discussed. For the interested reader they are given briefly in an appendix to this chapter.

The importance of the transitive property is that *any* two elements of the same sub-set in a partition are connected by the equivalence relation. And this is true whether the sorting is done by method one or by method two. If by method two, it follows directly from the transitive property. If by method one, we can always find an equivalence relation between any two elements of the same sub-set. For example:

Set	Partition	Equivalence relation
{coins in my pocket}	sub-sets of coins having the same value	has the same value as
{pots of paint}	sub-sets of pots containing paint of the same colour	is the same colour as
{novels in the library}	sub-sets of novels by the same author	is by the same author as
any set at all	any partition at all	is in the same sub-set as

Because of the close connection which it has with an equivalence relation, a sub-set belonging to a partition is called an *equivalence class*. To summarize so far: any equivalence relation, which can be applied to all elements of a given set, partitions the set into equivalence classes. And any partition of a set can be used to define an equivalence relation.

The interchangeability principle

Implicit in the idea of equivalence is that of interchangeability for a particular purpose. For paying a train fare, all coins in the 5p sub-set are interchangeable. For colouring my boat royal blue, all pots of paint in the royal blue sub-set are interchangeable. Someone who asks at the local library for 'any book by H. G. Wells' is thereby saying that it does not matter which particular one in that sub-set is chosen. This equivalence is with respect to a particular property only, the characteristic property of the equivalence class. So within the class, we can if we like choose a particular member because of its advantages in some different way. Within an equivalence class of coins of the same value, a collector might choose a coin for its mint condition. Within the equivalence class of royal blue paints, I would choose one which had good resistance to sun and salt water. The reader of H. G. Wells would probably prefer a book he or she had not already read.

A further consequence of the interchangeability principle is that it gives us another way of naming an equivalence class. The first way discussed was by its characteristic property (for example, 5p). A more concrete way is simply to use any member of the class as a representative (for example, a 5p coin). This is sometimes more convenient, but we must then be clear whether we are referring to the class thus represented or to the element itself. A representative does not, of course, by itself, define a class; for that we need to know also the equivalence relation in use. (Nor does an equivalence relation, by itself, define a class; we must also know the parent set.) To give a concrete example of the foregoing: a book by H. G. Wells could be shown as itself only; or, in a window display on his centenary, as a representative of his novels; or, in a lavish binding, as a representative of the output of a particular publisher (an equivalence class within the set of books produced by British publishers). So the method of naming a class by a representative works best when we have already well established our context and are using it as a convenient means of working within it.

Equivalence and equality

To say that two objects are equivalent means that they are alike in some particular way, which, if not obvious from the context, must be specified. To say that they are equal means that they are alike in every way. This can only be so if they are in fact the same object. And since an object can only be equal to itself, we might well expect statements of equality to be trivial. This is not necessarily the case.

This Rolls Royce = This Rolls Royce   is trivial, but
This Rolls Royce = My car   is not.

A statement of equality (called for short an *equation*) tells us that we are referring to the same object (which may be a physical object or an idea) in two different ways. Notice also that the statement 'This Rolls Royce' = 'My car' *cannot* be true, since the quotation marks mean that we are now referring to the words themselves, and not to the objects for which they stand. So

$$\text{three} \;=\; 3 \quad \text{is true, but}$$
$$\text{'three'} \;=\; \text{'3'} \quad \text{is false.}$$

If, however, we define an equivalence class by the characteristic property that all its members are names for the same number, then (using $\cong$ to mean *is equivalent to*)

$$\text{'three'} \;\cong\; \text{'3'} \quad \text{is true.}$$

If two *objects* are *equivalent*, then their *class properties* are *equal*. For example, since the books *The War of the Worlds* and *The Time Machine* are equivalent according to the relation defined earlier, their authors are the same person. If two coins are equivalent in value, the value of one is equal to the value of the other. If two objects are equivalent as defined by the relation that they balance when put in opposite scale pans, their weights are equal. So, if we sometimes regard objects as themselves and sometimes as representatives of their equivalence classes, in the first context they may be equivalent while in the second they (but really their classes) are equal. This may seem confusing, but, once the idea is grasped, it has the reverse effect, since it helps to make sense of otherwise confusing statements.

A second look at natural numbers

The foregoing ideas are of very general applicability, both everyday and mathematical. We were, in fact, already using them when we developed the idea of natural numbers. Here, the parent set was the set of all (finite) sets. The equivalence relation 'is in one-to-one correspondence with' partitions this set into equivalence classes, and *the characteristic properties of these classes are the natural numbers*. The last two sentences summarize what was said over several pages of Chapter 7, and it will be worthwhile at this stage turning back and comparing the two expositions.

When the ideas have been fully grasped, it can be seen that the second explanation really says everything that is essential – a good example of the condensation of mathematical thinking. As an initial explanation, the second would have been so condensed as to be for many people very hard to follow, but as a way of subsequently perceiving the essence of the matter, it is very effective.

Mathematical models

The second key idea in this chapter is that of a mathematical model.

Suppose that we are planning, say, a kitchen. It is helpful to make a scale plan of the room itself, and also to draw separately the various items of furniture which we intend to have in the room and cut these out. We can then try the latter in various positions and observe the result both for fit into the available space and for convenience. We have *abstracted* from the physical objects certain qualities in which we are interested for a particular purpose, in this case their overall size and shape as seen from above, and also their functions (cooker, refrigerator, table, etc.), while ignoring other qualities such as colour, cost, make. The qualities abstracted have been represented in the form of a paper model. This model embodies another important feature of the objects in the kitchen – that we can move them about. Moreover, if we arrive at an arrangement of the model in which all the items fit into a particular space, we know that the corresponding original objects – the cooker, refrigerator, etc. – will fit into the corresponding space in the kitchen. It is a working model.

Our model kitchen is a physical model, made for a particular purpose. The natural number system is a mental model, one of great versatility. But it uses the same basic method of abstracting, manipulating the abstractions instead of manipulating physical objects, and then re-embodying the result in the situation from which the abstractions were taken. In everyday life we do this habitually. We may be expecting a visit from friends. 'There are four of us, and they will be two grown-ups and three children.' The first stage of abstraction is that verbalized above, using primary concepts. For a particular purpose, say, laying the table, we are not interested in age, sex or whether resident or visitor. So we abstract still further: 4, 2, 3. In the general situation of having tea together, we concentrate on the combining aspect, and represent this by the mathematical operation of addition: $4 + 2 + 3$. Result: 9. First re-embodiment: there will be nine people. At this level of abstraction we match the set of people to the set of places, and then, physically, we lay nine places at the table.

The ability to think in this way we take for granted. But it is undeveloped in some primitive peoples – a traveller relates that having agreed on a price for a sheep, in terms of exchange articles, with a native, and having agreed also that two sheep were to be bought at that price, the seller would not accept two lots of articles and hand over two sheep. The first sheep was handed over in exchange for the specified articles, and the procedure then repeated. Although the seller had the concept of number, he could not form and manipulate the simple mathematical model appropriate to the situation. Trade on any scale is clearly dependent on this ability. Individual transactions may be done as above, but organized commerce requires the keeping of accounts, and the manipulation of this more developed model (as distinct from its formation) was greatly helped by the introduction of Hindu-Arabic notation, as has already been mentioned.

Measurement

One of the most striking things about the natural number system is the great variety of situations for which it provides a model. This is partly because the number of a set does not depend on what are the objects in the set or where they are. So the same numbers can be used as models for people, teacups, sheep, articles of trade, red blood cells (in a blood count), words (in a book) – for any collection of separate objects.

There are, however, certain situations for which numbers alone are not adequate. We cannot count the amount of milk in a bottle, or the length of a road, or the value of a car, or the hotness of an oven. But by combining a natural number with a unit of measure, we can extend its usefulness simultaneously in two different ways. We can use it for continuous quantities as well as for collections of discrete objects. And by varying our choice of unit we can make models for volume, length, value, temperature, weight, mass, area, time, velocity, electric potential, electric current, energy, frequency – the list could be continued to two or three times this length.

The basic principle involved in measurement is, of course, nearly as familiar to us as counting. Roughly speaking, we decide on a certain volume, weight, length, etc., and call this a *unit* of volume, weight, etc. We then find how many of these units must be put together to be equal to the weight, for example, of the object we are interested in. Thus we convert the question 'How much?', in the context of weight, to the question 'How many units of weight?' The answer is called a *measure* of the object's weight. Just as counting is a technique for finding the number of a set, so measurement is a technique for finding the measure of some particular quality of an object, for example, its volume, length, temperature.

In both counting and measurement, physical as well as mathematical activity is involved. With counting, the physical activity is usually slight and simple, such as pointing at or just looking in turn at the objects to be counted, unless the numbers are large or the objects to be counted are events happening in rapid succession (for example, counting the revolutions of an engine). For measurement we always need physical aids, for example, balances, rulers, liquid measures, thermometers. The physical side of the activity may be fairly straightforward, or complicated and requiring intricate apparatus. The latter are problems for the physicist and the instrument maker. Here we shall be centring our attention on the relation between the physical objects, the mathematical–physical activity of measurement, and the mathematical results of this activity. And as with counting, so we may expect to find with measurement, that there is more in it than meets the eye.

Weighing

Whereas there is only one kind of counting, there are many kinds of measuring. Since it is easier to think initially at the level of a typical example than at the next higher order of abstraction, we will begin by thinking about weighing.

Weight and mass are not, of course, the same thing. Weight is a force – the mutual attraction between the earth and an object. Mass is one way of describing the amount of matter in a body. So if a body is taken to the moon, its weight diminishes, but its mass remains the same. When we weigh something, it is sometimes because we do in fact want to know its weight (for example, if we are loading a ship or an aeroplane), but very often because we want to know its mass – how much of the stuff we are getting (for example, if we are buying cheese or coal). Weighing is a very convenient way of measuring mass, because bodies of equal mass, at the same place, have equal weights. So, for our present purposes, it does not matter whether it is the mass of a body or its weight that we are most interested in. A pair of scales, or balance, is a device whereby we can compare the weight of two bodies. It tells us whether or not they weigh the same, and if not, which is the heavier.

So we can choose any object we like for our standard mass – the internationally accepted one is a kilogram (kg.), kept by the Bureau International des Poids et Mesures, near Paris – and by this matching process prepare a set of objects (say, lumps of iron) all having the same weight as the international standard kilogram. These are usually called 'kilogram weights', but, as a reminder that weight is a force, not, for example, a lump of iron, we shall call them 'kilogram objects'.

We can combine the weights of several of these kg. objects by putting them in the same scale pan. If the scales exactly balance with a bag of flour in one pan and, say, five kg. objects in the other, then we say that the weight of the bag of flour is 5 kg. For convenience we can also, by this method, prepare a set of standard objects whose measures of weight are 1 kg., 2 kg., 4 kg., etc. If the scales now balance with a bag of potatoes in one pan and a 2 kg. and a 4 kg. object in the other, we say that the weight of the potatoes is 6 kg. Implicit in this is an assumption that adding these units is a true model for combining gravitational forces. This happens to be the case, but it should not be taken for granted. One litre of water at a temperature of 10 °C, combined (by pouring into the same container and stirring) with one litre of water at a temperature of 40 °C, gives two litres of water, but not at a temperature of 50 °C. A journey of length 10 kilometres travelled at 40 kilometres per hour, combined (by starting one where the other finishes) with a journey of length 5 km. travelled at 60 km./h results in a journey of length 15 km., but not at 100 km./h. This warns us that in more complex cases, where we not only add but multiply, factorize, solve equations and manipulate the mathematical model in increasingly complex ways, we must not do so unthinkingly.

Three realms of thought

So let us conclude this chapter by distinguishing between three realms of thought which are involved, and the ways in which they correspond.

Realm 1: physical objects, events or other observations.

For example, (a) Some heavy books we want to take on a journey by air.

(b) Water coming out of hot and cold taps for our bath.

(c) Six electric unit cells for our tape recorder.

Realm 2: physical qualities of these objects.

For example, (a) Their weights.

(b) Their temperatures.

(c) Their electro-motive forces.*

Realm 3: mathematical ideas (in this case measures of these qualities).

For example, (a) Measures of weight – numbers of kilograms or grams.

(b) Measures of temperatures – numbers of degrees Celsius.

(c) Measures of e.m.f. – numbers of volts.

In each of these realms, there are corresponding operations of which the results must also correspond if the model is to be a successful one.

Realm 1: operations on (mental representations of) *physical objects.*

For example, (a) Packing all in the same suitcase.

(b) Running both into the same bath and mixing.

(c) Connecting these cells in series.

Realm 2: operations on their physical qualities.

For example, (a) Combining their weights.

(b) Combining their temperatures.

(c) Combining their e.m.f.s.

* Electro-motive force, commonly called voltage, may conveniently be abbreviated to e.m.f.

Realm 3: mathematical operations.

For example, (a) Adding the numbers of kilograms or grams.
                (b) Insufficient data – we need to know also the rates of
                      flow – but certainly not adding.
                (c) adding the numbers of volts.

Examples like (a) and (c) indicate how various are the physical realities for which virtually the same mathematical model can be used, simply adding numbers of units. Examples like (b) remind us to be careful; but even in this case the appropriate model, if the rates of flow are equal, is addition followed by one more operation (division by 2). If they are not equal, the model for combining is represented by the formula $(f_1 t_1 + f_2 t_2) \div (f_1 + f_2)$, where f_1 and f_2 represent the two rates of flow in litres per minute and t_1 and t_2 the two temperatures.

Whenever someone discovers by what combination of mathematical operations they can successfully predict some physical result, they have discovered a new mathematical model. And whenever we use mathematics to help us in our everyday or scientific activities, whether we are adding the items of a bill or calculating the resonant frequency of an electric circuit, we do so by making and manipulating mathematical models.

APPENDIX
Properties of an equivalence relation

Let R stand for any relation and let a, b, c, be any members of the set of objects for which this relation is defined. Then R is an equivalence relation iff, whichever objects we choose,

  (i) $a R b$ and $b R c$ implies $a R c$
 (ii) $a R a$
(iii) $a R b$ implies $b R a$

In words, an equivalence relation is (i) transitive, (ii) reflexive, (iii) symmetric.

CHAPTER 10

A Need for New Numbers

A problem always arises with measurement which is absent from counting, that of a mis-match. Given any (finite) set of separate objects, we can always find the number of this set exactly; which is to say that we can always match it exactly with a set of counting words. But when measuring the weight, length, volume, temperature ... of a body by matching it in some way with the same quality in another physical object, we can never be sure of finding a standard object which gives a perfect match. In the weight example, with a unit as large as the kilogram, the chance that a given object will balance exactly with some number of kilogram objects is small. So our mathematical model cannot always represent the physical object accurately.

A simple way of dealing with this is to use smaller units. Where this is not convenient, then we must cut up our standard object into smaller equal parts, and hope to find a combination of these which will match the given object. If we are measuring weight, this match will be by balance. If we are measuring length, the match will be by putting side by side. If we are measuring time, we will start and stop the standard event (for example, movement of a stop-watch) simultaneously with the start and finish of the event whose duration we wish to measure. Given objects and standard objects are matched in different ways for measuring different qualities, but the mathematical problem is the same in all – to construct a model for these cut-up and combined units, which are called *fractional* units. As in the natural number system, we want a working model. So combining physical qualities must be represented by an appropriate operation in the model. If we call this operation 'adding', we must be clear that we are not implying that it is exactly the same operation as adding natural numbers; and if we call the elements in our new model '*fractional numbers*', we must be equally clear that these may or may not have the same properties as natural numbers. We use the same words as for natural numbers in a way which at this stage is perhaps premature. We do so because we are hoping to generalize the corresponding ideas, in the manner described in Chapter 3, page 57.

It will be a great advantage if we can arrive at a notation for fractional numbers which (i) is based on the same numerals as are used for natural numbers, (ii) allows us to use the same methods for adding as those which we have learnt for natural numbers, either as they stand, or extending them by learning a few extra procedures, as we did when we developed our short-multiplication schema into long multiplication.

Fractions

The model we are going to develop will be the same, except for the nature of the units, for all the physical qualities for which we are using it. So when we refer to 'cutting up', we are using it in the generalized sense of any way of breaking up into parts. (We want to keep 'division' for the mathematical operation: compare 'uniting' sets and 'adding' numbers.)

This represents any standard object,

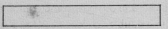

and this represents it cut up into five.

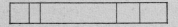

Clearly this way of cutting up is of no use for measurement, however, since we do not know how big* the bits are; and whether or not we get a match with our given object will depend on which bit we choose.

If we cut up our standard object into bits which match each other, according to whatever quality we are trying to measure, this gets over the second problem, of which one we choose. How big the bits are will then depend on how many of them there are. A cutting-up of this kind we will call *sharing*; bits which are equivalent (that is, which match in the way described) we will call *parts*; and we will describe the size of the parts by saying into how many of these parts we have shared our

---

* In the generalized sense; that is, how heavy, how long, how great an e.m.f., etc.

standard object. So this represents a standard object * shared into *fifth parts*.

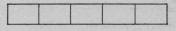

This, *eighth parts*.

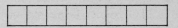

This, *third parts*.

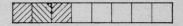

This represents the result of sharing into eighth parts, and then combining † three of these parts.

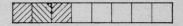

We call this fractional part of an object 'three eighth parts' of the object, or for short 'three eighths' of it. A fractional part is thus a part which is obtained by a double action of sharing and combining. Abstracting what is common to all these double actions, we get in Realm 3 a mathematical double operation which is called a *fraction*.

The mathematical notation for this double operation is $\frac{3}{8}$ (read this as 'three over eight'). Since the numeral below the line tells us the name of the parts represented – whether they are fifth parts, eighth parts, third parts, etc. – this is called the denominator of the fraction. The numeral above the line tells us how many such parts are combined, and is called the numerator.

The notation $\frac{3}{8}$ might appear to suggest, both from our habit of reading downwards and from its often being written as 3/8 for convenience in printing or typing, that the combining is done first, whereas in the foregoing description we first share into 8 parts and then combine 3

* Though it has been emphasized that this is a generalized object on which generalized actions of sharing and combining are to be done, it is nevertheless a help to one's thinking to imagine it as something more concrete, such as a cake which we are literally cutting up into fair shares – that is, equal volumes of cake.

† In whatever way is appropriate to the physical quality concerned: if weight, by putting in the same scale pan; if volume of cake, the same person eats it.

of these. However, we shall see that these actions are commutative – we get the same result whichever we do first. So the notation $\frac{3}{8}$ may be taken as representing simultaneously *both* of the two possible orders of the mathematical double operation.

Start with a standard object.

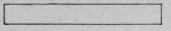

Share into 8 parts.

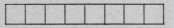

Combine 3 of these eighth parts: result, three eighth-parts of an object.

Now the other way about. Start with a standard object.

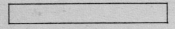

Combine 3 of these standard objects.

Share this into 8 parts: result, one eighth part of three objects.

Except for their arrangement (which does not affect the quantity), the shaded part is the same as before. So the fraction $\frac{3}{8}$ represents ($\div$ 8 $\times$ 3), as embodied in the first set of diagrams above, and ($\times$ 3 $\div$ 8), as embodied in the second set of diagrams. This is one reason for reading $\frac{3}{8}$ as 'three over eight', rather than 'three eighths', which implies only the first of these alternative orders.

Equivalent fractions

We shall now reverse the process. By using embodiments of these double operations which we call fractions, we can find sets of equivalent fractions, and an equivalence relation between fractions.

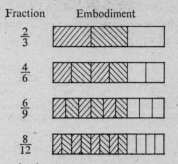

and so on; the pattern is clear.

Though the fractions themselves are different, they correspond to the same amounts of whatever physical quality we are concerned with. If we applied the corresponding actions of sharing and combining to a standard object, the resulting part-objects would match. With units attached, the fractions $\frac{2}{3}$, $\frac{4}{6}$, $\frac{6}{9}$, $\frac{8}{12}$... represent equal measures. (Equal amounts of cake, in the concrete example.) In this respect they are therefore equivalent, and we may collect them together into the equivalence class $\{\frac{2}{3}, \frac{4}{6}, \frac{6}{9}, \frac{8}{12}...\}$

In the same way we can find other sets of equivalent fractions. For example:

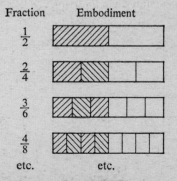

Set of equivalent fractions: $\{\frac{1}{2}, \frac{2}{4}, \frac{3}{6}, \frac{4}{8}...\}$

Another example, this time without diagrams: $\{\frac{5}{8}, \frac{10}{16}, \frac{15}{24}, \frac{20}{32} \cdots\}$
Not only is the pattern of each equivalence class clear, but a general method for forming them is beginning to emerge.

Start with any fraction, $\frac{9}{12}$
double both upper and lower numbers, $\frac{18}{24}$
treble both upper and lower numbers, $\frac{27}{36}$
etc.
Equivalence class $\{\frac{9}{12}, \frac{18}{24}, \frac{27}{36} \cdots\}$
And, in general, if a, b, k, are natural numbers, then

the fraction $\frac{a}{b} \cong$ (is equivalent to) $\frac{ka}{kb}$

This works both ways: since also $\frac{ka}{kb} \cong \frac{a}{b}$, we can get another fraction equivalent to any given fraction by either multiplying or dividing the numerator by the same natural number. The former we can always do; the latter, which is the well-known 'cancelling' rule, sometimes.

Example: $\dfrac{9}{12} = \dfrac{3 \times 3}{4 \times 3} \cong \dfrac{3}{4}$

We also have: $\dfrac{3}{4} \cong \dfrac{3 \times 2}{4 \times 2} = \dfrac{6}{8}$

So these also belong to the last named equivalence class, which

we can now write: $\left\{\dfrac{3}{4}, \dfrac{6}{8}, \dfrac{9}{12}, \dfrac{18}{24}, \dfrac{27}{36} \cdots\right\}$

Fractional numbers

The characteristic property of any set of equivalent fractions we call a *fractional number*.* With a unit attached, each fraction in an equivalence

* Compare with natural numbers, page 166.

class represents the same measure, and, without the unit, it represents the same number. This means that we can use any fraction from the set as a name for the number of that set; and, although this invites confusion if we do not know what is going on, if we do know, it has considerable advantages for purposes of calculation.

So if we are talking about fractions, which are double operations,

$$\frac{2}{3} \cong \frac{4}{6}$$

If we are talking about fractional numbers,

$$\frac{2}{3} = \frac{4}{6}$$

for each denotes the same equivalence class. The sign in the middle therefore indicates which of the two is meant.

Adding fractional numbers. We want this mathematical operation to correspond to combining part-objects. This is straightforward if the numbers are represented by fractions having the same denominator, for we are then combining part-objects of the same kind, an essential already noted on page 174. But we have to remember that adding does not mean quite the same for fractional numbers as for natural numbers. To remind ourselves of this we use $\oplus$ for the new kind of addition, and $+$ for the old kind.

Example:

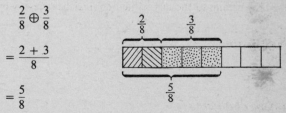

$$\frac{2}{8} \oplus \frac{3}{8}$$

$$= \frac{2 + 3}{8}$$

$$= \frac{5}{8}$$

If the denominators are not equal, this is where the interchangeability principle within equivalence sets (page 164) comes to our help. Since all the fractions in an equivalence set stand for the same number, we can choose whichever ones suit us best for some other purpose, in this case a calculation.

Suppose that we want to add, say, $\dfrac{2}{4} \ \oplus \ \dfrac{3}{9}$

Replace by these equivalent fractions $\dfrac{2 \times 9}{4 \times 9} \oplus \dfrac{3 \times 4}{9 \times 4}$

which stand for the same numbers $\dfrac{18}{36} \ \oplus \ \dfrac{12}{36}$

as before. For denominator, we choose

$4 \times 9 = 36$

Now we can add. $\dfrac{18 \ + \ 12}{36}$

$= \ \dfrac{30}{36}$

It should, of course, make no difference which fractions we use as replacements, provided that they stand for the original numbers and have the same denominators. Let us try the calculation by a different route.

First we will replace $\dfrac{2}{4} \ \oplus \ \dfrac{3}{9}$

the original fractions by $= \dfrac{1 \times 2}{2 \times 2} \oplus \dfrac{1 \times 3}{3 \times 3}$

equivalent ones using
the cancelling rule. $= \ \dfrac{1}{2} \ \oplus \ \dfrac{1}{3}$

Now we can find a $= \dfrac{1 \times 3}{2 \times 3} \oplus \dfrac{1 \times 2}{3 \times 2}$

smaller common $= \ \dfrac{3}{6} \ \oplus \ \dfrac{2}{6}$

denominator, namely $= \ \dfrac{3 + 2}{6}$

$2 \times 3 = 6$ $= \ \dfrac{5}{6}$

This answer looks different, but, of course, $\dfrac{5}{6}$ represents the same fractional number as $\dfrac{30}{36}$, since $\dfrac{30}{36} = \dfrac{5 \times 6}{6 \times 6} = \dfrac{5}{6}$. So we have verified that the interchangeability principle works in this case. A general proof is not difficult, but requires the use of algebra.

Multiplying fractional numbers. As yet we have no meaning for 'multiplying' in the new context of fractional numbers. We could, of course, decide to do without a meaning – there are plenty of mathematical systems which have only one operation. But we shall then not have generalized the natural number system completely, so we ought to try. We can either look for a meaning for 'multiplication' which is satisfactory in the realm of pure mathematics, and then see whether it provides a useful working model for Realm 1, or we can use the requirement of a satisfactory working model to suggest a meaning, and then check whether it is mathematically acceptable. Both approaches have their merits. The latter, being less abstract, is the one we shall use here.

Start as usual with a standard object.

Then this object represents the fraction $\frac{2}{3}$.

In natural numbers, 3×4, when embodied in physical objects, means: start with a three-set

and combine 4 of these.

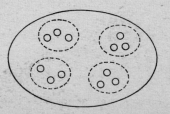

So, in fractional numbers, $\frac{2}{3} \otimes \frac{4}{5}$ might reasonably mean: start with two third-parts of an object,

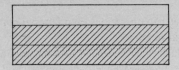

and take four fifth-parts of this.

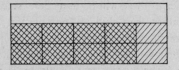

In natural numbers, 'calculate 3×4' means 'find the number of the resulting set'. In fractional numbers, 'calculate $\frac{2}{3} \times \frac{4}{5}$' might therefore reasonably mean 'find what fractional part of the object the resulting part-object is'. The resulting part-object is shown by the cross-hatched area. The original object has now been shared into 15 parts (3×5), and the cross-hatched area combines 8 (2×4) of these.

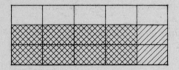

This suggests that $$\frac{2}{3} \otimes \frac{4}{5} = \frac{2 \times 4}{3 \times 5} = \frac{8}{15}$$

would be a reasonable way to multiply these fractions – reasonable, in the sense that it gives a good working model for part-objects of part-objects. It also satisfies requirements (i) and (ii) on page 174 very well.

These two methods, for addition and multiplication of fractional numbers, are, of course, those which have been agreed by mathematicians – we have been pretending we did not know in order to try to see how they were arrived at. Stated generally, if a, b, c, d, are natural numbers, then the method for adding is:

$$\frac{a}{d} \oplus \frac{b}{d} = \frac{a + b}{d}$$

and the method for multiplying is

$$\frac{a}{b} \otimes \frac{c}{d} = \frac{a \times c}{b \times d}$$

where $\oplus$ and $\otimes$ refer to operations on fractional numbers and $+$ and $\times$ to those on natural numbers.

There is still much unsaid about fractional numbers. Techniques for manipulating them have not been systematized, and decimal notation – which can greatly simplify some of these manipulations – has not been introduced. Neither of these will be done here, since the present aim is comprehension rather than. skill at computation. Also, we have not checked that the fractional numbers have the five properties of the natural number system which we found in Chapter 8 to be so important. This we must certainly do. Since the treatment is algebraic, it has been put into an appendix to this chapter. The reader who does not think easily in algebraic terms may take it on trust, since he already has the ideas and only requires to be assured that they also hold good for fractional numbers. There is also a third matter of importance, which is whether and to what extent natural and fractional numbers can be intermixed. This last point will be discussed in Chapter 11, with the help of the ideas of isomorphism and mathematical generalization.

APPENDIX

Fractional numbers have the five properties of the natural number system

Let a, b, c, d, e, f, x, y, stand for any natural numbers.

Then $\frac{a}{b}$, etc., will represent fractional numbers.

ADDITION IS COMMUTATIVE

We can only add if the denominators are equal.

$$\frac{a}{d} \oplus \frac{b}{d} = \frac{a + b}{d} = \frac{b + a}{d} = \frac{b}{d} \oplus \frac{a}{d}$$

This property follows immediately from the corresponding property for natural numbers, and the same is true for all the other properties.

ADDITION IS ASSOCIATIVE

$$\left(\frac{a}{d} \oplus \frac{b}{d}\right) \oplus \frac{c}{d} = \frac{(a+b)}{d} \oplus \frac{c}{d} = \frac{(a+b)+c}{d} = \frac{a+(b+c)}{d}$$

$$= \frac{a}{d} \oplus \frac{(b+c)}{d} = \frac{a}{d} \oplus \left(\frac{b}{d} \oplus \frac{c}{d}\right)$$

MULTIPLICATION IS COMMUTATIVE

$$\frac{a}{b} \otimes \frac{c}{d} = \frac{a \times c}{b \times d} = \frac{c \times a}{d \times b} = \frac{c}{d} \otimes \frac{a}{b}$$

MULTIPLICATION IS ASSOCIATIVE

$$\left(\frac{a}{b} \otimes \frac{c}{d}\right) \otimes \frac{e}{f} = \frac{a \times c}{b \times d} \otimes \frac{e}{f} = \frac{(a \times c) \times e}{(b \times d) \times f}$$

$$= \frac{a \times (c \times e)}{b \times (d \times f)} = \frac{a}{b} \otimes \frac{c \times e}{d \times f} = \frac{a}{b} \otimes \frac{c}{d} \otimes \frac{e}{f}$$

MULTIPLICATION IS DISTRIBUTIVE OVER ADDITION

$$\frac{x}{y} \otimes \left(\frac{a}{d} \oplus \frac{b}{d}\right) = \frac{x}{y} \otimes \frac{(a+b)}{d}$$

$$= \frac{x \times (a+b)}{y \times d}$$

$$= \frac{x \times a + x \times b}{y \times d}$$

$$= \frac{x \times a}{y \times d} \oplus \frac{x \times b}{y \times d}$$

$$= \left(\frac{x}{y} \otimes \frac{a}{d}\right) \oplus \left(\frac{x}{y} \otimes \frac{b}{d}\right)$$

CHAPTER 11

Further Expansions of the Number Schema

Knowing in advance that further examples of the concept *number system* are on the way, it is time for us to decide on a characteristic property, or set of properties, whereby we can be sure whether any particular mathematical system which we encounter is in fact a number system. It will probably come as no surprise that the five properties which were found in Chapter 8 to be of such importance for calculation with the natural numbers, and which in the appendix to Chapter 10 we found also to hold true for the fractional numbers, have been agreed among mathematicians to be the characteristic properties of a number system. The following definition also makes explicit some further requirements which we have so far taken for granted.

Definition. A number system is a set of mathematical concepts, called numbers, together with two closed binary operations (see below) on the set, called addition and multiplication, such that

> addition is commutative and associative
> multiplication is commutative and associative
> multiplication is distributive over addition

Notice that the definition does not specify whether the set of numbers is finite or infinite. We can in fact find systems which satisfy the above definition containing as few as two numbers. The number systems to be introduced in this chapter are, however, like the natural and the fractional numbers, infinite.

A binary operation is an operation which combines two mathematical ideas to get a third, addition and multiplication being familiar examples. 'On the set' means that the binary operation can be applied to any two elements of the set, and 'closed' means that the third idea thus obtained also belongs to the set. These requirements are non-trivial – subtraction, on the set of natural numbers, is not closed, and neither is division, which means that, as long as we are confined to this system, there are many problems in Realm 1 for which we have no mathematical model. (Examples: I have £100 in the bank and withdraw £150; how much have I left? Share 10 apples between 3 people. Although there are answers to both of these problems, they can only be obtained by going outside the natural number system.)

In this chapter we shall see how the original number schema can be further expanded to include three new number systems: integers, rational numbers and real numbers. Clearly the treatment cannot be so thorough as it was for the two earlier systems, the aim being to give an overview of the development of these new systems, in terms of the mathematical and psychological ideas already established, particularly those of expansion, restructuring and mathematical generalization.

Before we embark on this, there is a further aspect of the progress from natural to fractional numbers which is of considerable practical importance.

Can we mix numbers from different systems?

When, in the appendix to Chapter 10, we examined whether fractional numbers formed a number system according to our present definition, we found not only that they did but that the required properties were directly transmitted from the natural to the fractional numbers by the way in which the latter, and $\oplus$ and $\otimes$ on the latter, were defined. The fractional notation, too, played an important part in enabling us to extend our well-learnt routines for operating on the natural numbers for use with the new system. As an example of expansion it could hardly be bettered, and the question now arises, is the assimilation so complete that the new and the old systems can be freely intermixed? Alternatively, since we already know that it is in fact common practice to do so, we could ask to what extent this is valid mathematically – whether, so to speak, it is good grammar, or just a convenient slang.

Certain fractional numbers already appear to have a foot in both camps. These are $\frac{1}{1}$, $\frac{2}{1}$, $\frac{3}{1}$ ... Since each of these represents an equivalence class of fractions (for example, $\frac{3}{1}$ represents $\{\frac{3}{1}, \frac{6}{2}, \frac{9}{3} ...\}$), it would be hard to justify *equating* these with the natural numbers 1, 2, 3 ... But it is entirely legitimate to note the correspondences

$$\tfrac{1}{1} \leftrightarrow 1, \tfrac{2}{1} \leftrightarrow 2, \tfrac{3}{1} \leftrightarrow 3 \ldots$$

and to ask how far this correspondence also holds good between the two basic operations on these two sets.

Natural number system	Fractional number system
$3 + 2 = 5$	$\tfrac{3}{1} \oplus \tfrac{2}{1} = \tfrac{5}{1}$
$3 \times 2 = 6$	$\tfrac{3}{1} \otimes \tfrac{2}{1} = \tfrac{6}{1}$

And, in general, if

$$a \leftrightarrow a$$
$$b \leftrightarrow \beta$$

then $\qquad\qquad a + b \leftrightarrow a \oplus \beta$

and $\qquad\qquad a \times b \leftrightarrow a \otimes \beta$

This further degree of correspondence is expressed by saying that the natural numbers together with + and × are *isomorphic* (of the same form) with this particular sub-set of the fractional numbers together with $\oplus$, $\otimes$. In general an isomorphism means that if we do corresponding operations on corresponding elements of the two sets concerned, then the results also correspond. In the present context it allows us to work in whichever system is the more convenient, provided that we use the methods for adding and multiplying which are appropriate to the system.

We cannot calculate, say, $\qquad 2 + \tfrac{1}{2}$

since we have no way of adding natural and fractional numbers.

But if we replace this by $\qquad \tfrac{4}{2} \oplus \tfrac{1}{2}$

the result is $\qquad\qquad \tfrac{5}{2}$

Conversely, if we write $\qquad \tfrac{10}{3}$

as $\qquad\qquad\qquad \tfrac{9}{3} \oplus \tfrac{1}{3}$

and replace this by $\qquad 3 + \tfrac{1}{3}$

this tells us that, in Realm 1, taking 10 apples and sharing these into 3 parts may be done by giving everyone 3 whole apples and 1 third-part of an apple.

We can add $\qquad\qquad 2 + \tfrac{1}{2}$ and $3 + \tfrac{1}{3}$

either by working entirely in the fractional-number system:

$$\frac{12}{6} \oplus \frac{3}{6} \oplus \frac{18}{6} \oplus \frac{2}{6}$$
$$= \frac{35}{6}$$

or by collecting together numbers of the same systems

$$2 + 3 + \frac{1}{2} \oplus \frac{1}{3}$$
$$= \quad 5 + \frac{5}{6}$$

The reader may care to verify that the mixed number $5 + \frac{5}{6}$ corresponds to the fractional number $\frac{35}{6}$. It is this isomorphism between the natural-number system and a sub-set of the fractional-number system that has, without our knowing of its existence, taken care of our past incautious mixings of the number systems, in somewhat the same way as gravity kept us safely on the ground long before we were consciously aware of it. So, when we explore further number systems, isomorphism is another property which, though not indispensable to a number system as such, is necessary if we want to go on mixing our systems.

When working with mixed numbers in this way, it is convenient and usual to use the signs $+$, $\times$, for addition and multiplication in both systems, interpreting them according to their immediate context. The justification for this is not, however, its convenience, but the isomorphism which allows us to work with mixed number systems, together with an awareness of the two meanings which each symbol may have.

Opposites which cancel

Compare the results of combining these two kinds of opposites.

First kind.
Two large apples and two small apples.
Three fat men and three thin men.
Five black sheep and five white sheep.

Second kind.
Walking up two steps and walking down two steps.
Temperature falling by 3°C and rising by 3°C.

Paying five pounds into the bank and withdrawing five pounds from the bank.

The first three results are: four apples, six men, ten sheep. The second three results are: stay still, no change in temperature, no change in bank balance. The mathematical models for the first three are, in the natural number system: $2 + 2 = 4, 3 + 3 = 6, 5 + 5 = 10$. But if we are going to keep addition for our mathematical model of combining, we shall need a different kind of numbers for the second set of examples. These are all combinations not of sets of physical objects but of reversible events, such that a combination of two equal and opposite events has the same end-result as no event. So 'combine' has a different meaning in these two contexts.

Integers

In the mathematical realm, adding two and subtracting two are opposite operations which cancel. So let us represent these operations by $(+2)$ and (-2), using parentheses here to show that the $+$ or $-$ sign is fused with the 2 to represent a new kind of number. The binary operation $\oplus$ on these we will define by the result.

$$(+2) \oplus (-2) = 0$$

This provides a satisfactory model for the first example of the second kind, and

$$(-3) \oplus (+3) = 0$$
$$(+5) \oplus (-5) = 0$$

will do equally well for the second and third.

These new numbers are called *integers*. The use of $\oplus$ for their addition means, as before, that it both resembles and differs from the addition of natural numbers. It does not, however, identify the addition of integers with that of fractional numbers; a new sign might be useful, but is not available without casting special type.

With this as a start, let us try to develop a new number system, basing our ideas initially on events in Realm 1, and checking afterwards that they fulfil the mathematical requirements already established.

We need a typical example: the first of the second kind on page 188 will do.

Physical realm	Mathematical model
Go up 2 steps and go up 3 steps	$(+2) \oplus (+3) = (+5)$
Go down 2 steps and go down 3 steps	$(-2) \oplus (-3) = (-5)$
Go up 2 steps and go down 3 steps	$(+2) \oplus (-3) = (-1)$
Go down 2 steps and go up 3 steps	$(-2) \oplus (+3) = (+1)$

The method for adding integers can be seen to be straightforward, being based directly on this or any other suitable physical embodiment. Multiplication, however, presents a problem, since it is hard to think of a meaning for, say, 'Go up 2 steps times go down 3 steps.' The difficulty lies in the fact that both $(+2)$ and (-3) represent operations; so we have to look for a guide to our thinking about operations *on* operations.

In the physical realm we can do an act, and (sometimes) we can undo it. So let us represent

Do once   by $(+1)$
Do twice   by $(+2)$, etc., and
Undo once   by (-1)
Undo twice   by (-2), etc.

This is consistent with our existing work on addition. For example,
Do twice and undo three times
has the same result as undo once. $(+2) \oplus (-3) = (-1)$

But it has the advantage that we can use it in a context which represents operations on operations. For example,
Do twice (go down 3 steps)
has the same results as
(go down 6 steps). $(+2) \otimes (-3) = (-6)$

To undo an act, we have to do an equal and opposite act.
Undo once (go down 3 steps)
means the same as
(go up 3 steps). $(-1) \otimes (-3) = (+3)$

Common sense also gives the following results in the physical realm, suggesting the corresponding mathematical results given on the right.

Undo twice (go down 3 steps)
has the same results as
(go up 6 steps). $(-2) \otimes (-3) = (+6)$
Do twice (go down 3 steps)
has the same result as
(go down 6 steps). $(+2) \otimes (-3) = (-6)$

You will find it useful to invent further examples for yourself, and formulate explicitly a general method for multiplying integers.

Have we constructed a number system? *

It is now time to see whether the integers, abstracted from quantifiable reversible events, together with the operations $\oplus$ and $\otimes$, developed as above, are a number system.

That addition of integers is associative and commutative can be seen directly from the physical events from which the method was derived. That multiplication is commutative is non-trivial, since the second integer represents an act or event, while the first represents a doing or undoing of this act. However, we can easily verify that, for example,

Do twice (go down 3 steps) $(+2) \otimes (-3)$
has the same result as
Undo 3 times (go up 2 steps): $= (-3) \otimes (+2)$
in each case
(go down 6 steps). $= (-6)$

The general method for multiplying integers is given by the following four statements, in which a, b, are natural numbers, while $(+a)$, $(-a)$, $(+b)$, $(-b)$, are integers. At this stage it is convenient to introduce the widely used abbreviation ab for $a \times b$, to avoid having parentheses within parentheses.

$$(+a) \otimes (+b) = (+ab)$$
$$(+a) \otimes (-b) = (-ab)$$
$$(-a) \otimes (+b) = (-ab)$$
$$(-a) \otimes (-b) = (+ab)$$

---

* This section may be omitted without loss of continuity.

That multiplication of integers is commutative follows directly from the above. There are four cases to consider. Here is one:

$$(+a) \otimes (-b) = (-ab)$$
$$(-b) \otimes (+a) = (-ba) = (-ab)$$

whence *
$$(+a) \otimes (-b) = (-b) \otimes (+a)$$

As with fractions, the property of integers follows directly from that of natural numbers if we agree to define $\otimes$ on integers as above. The same is true for the associativity of multiplication, which is more easily understood in the abstract than in a physical embodiment. Readers may care to construct their own proof.

A general proof that multiplication is distributive over addition is cumbersome, because of the many possible cases. Here is one of them:

$$(-a) \oplus (-b) = [-(a + b)] \dagger$$

whence

$$(-n) \otimes [(-a) \oplus (-b)] = (-n) \otimes [-(a + b)]$$
$$= \{+[n \times (a + b)]\}$$
$$= [+(na + nb)]$$

Also

$$(-n) \otimes (-a) \oplus (-n) \otimes (-b) = (+na) \oplus (+nb)$$
$$= [+(na + nb)]$$

Hence

$$(-n) \otimes [(-a) \oplus (-b)] = (-n) \otimes (-a) \oplus (-n) \otimes (-b)$$

The above looks more difficult than it really is, primarily because the signs $+$, $-$, are doing double duty. They stand for addition and subtraction of natural numbers, and also, when written thus $(+a)$, $(-a)$, to distinguish between positive and negative integers. This breaks the rule, suggested in Chapter 4, that within the same context symbols should have one meaning only. My only excuse for doing so is that this notation is the one most commonly used by mathematicians.

Here is an alternative notation which, though not in general use, has been found helpful in getting the ideas clear. It is based on the familiar

---

* Notice that we have used both the transitive and the symmetric property of equality.

† Here we need two kinds of brackets. The parentheses () mean that we are thinking of $(a + b)$ as a single natural number, while the square brackets [] enclose the minus sign and the natural number $(a + b)$ to represent a negative integer.

number line. Opposite actions are represented first by movements in opposite directions along a line, for example,

Go up 2 steps and

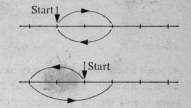

go down 2 steps.

Go down 2 steps and

go up 3 steps.

These movements are represented in turn by R and L for the direction and natural numbers for the number of units of distance moved. The above now becomes

$$R2 \oplus L2 = 0$$
$$L2 \oplus R3 = R1$$

Addition of integers in this notation is very straightforward, particularly since the visual aid of movements along a line is always available. For multiplication, we can use D for 'do' and U for 'undo', so, for example,

Do twice (go left 3 steps)	$D2 \otimes L3$
means the same as	
go left 6 steps.	$= L6$
Undo twice (go left 3 steps)	$U2 \otimes L3$
means the same as	$= R6$
go right 6 steps.	

All the earlier work in this section can if desired be translated into this notation. Readers who wish to consolidate their ideas about integers may care to experiment for themselves.

Mixing integers and natural numbers

There is clearly an isomorphism between the positive integers together with zero, and the natural numbers. So it is a common practice to write positive integers simply as 1, 2, 3, etc., dropping the + sign.

It is certainly more convenient to write $2 + 3 = 5$
than $(+2) \oplus (+3) = (+5)$

and since the result is the same in both number systems, it is reasonable to work in the simpler one. But what do we mean by this?

$$2 - 5 = -3$$

This subtraction cannot be done in natural numbers, so it must mean either

$$(+2) \ominus (+5) = (-3)$$
or $$(+2) \oplus (-5) = (-3)$$

We have not discussed subtraction of integers, for the same reason as division by a fraction was omitted – both involve the idea of an inverse, which is necessary for completeness but not for an overview. Subtracting (-5) is in fact defined as adding $(+5)$, so we get the same result whichever of the two above meanings we choose. Nevertheless, the statement

$$2 - 5 = -3$$

is an ambiguous one, and still more ambiguous are those like

$$4 \times (2 - 5) = 4 \times (-3) = -12$$

An accurate statement of the latter is

$$(+4) \otimes [(+2) + (-5)] = (+4) \otimes (-3) = (-12)$$

We get the right result from calculations of the earlier kind because the integers behave so similarly to the natural numbers that we can mix the two systems rather freely. But 'mix' can mean either 'intermingle' or 'confuse', and when as beginners we mix natural numbers and integers, it is more often with the latter meaning. However, the mathematical structure usually saves us from going wrong; and when we come to look more closely at the pitfalls from which it has saved us, some of us may decide that ignorance was bliss!

Rational numbers

Like natural numbers and integers, rational and fractional numbers are widely confused, and largely for the same reason: they behave so alike that we can get away with it. Their origins in Realm 1 are however quite different. Fractional numbers are properties of equivalence classes of fractions, which are abstracted from actions of collecting and sharing. Rational numbers are properties of equivalence classes of ratios, which are abstracted from a certain kind of correspondence between sets. Consider the following tables.

The first gives distances walked by a hiker in different times.

Distance from start in kilometres	5	10	15...
Time in hours	2	4	6...

The second gives the volume of concrete spread in one hour by varying numbers of labourers.

Volume in cubic metres	5	10	15...
Number of men	2	4	6...

The third refers to the height above the starting point of a sloping path.

Distance along path in metres	5	10	15...
Height in metres	2	4	6...

The same mathematical model, with the units omitted, serves for these three different physical situations, and for many others. Its essential feature is the way in which the numbers from the two sets $\{5, 10, 15 \ldots\}$ and $\{2, 4, 6 \ldots\}$ are paired, thus:

$$5:2, \qquad 10:4, \qquad 15:6 \ldots$$

These are read as '5 to 2', '10 to 4', etc., meaning roughly 'from 5 in the first set go to 2 in the second set', etc. (This meaning will be developed further in Chapter 13, in the section on *Functions*.) Each of these pairs of numbers is called a *ratio*, and we shall now develop the idea of an equivalence class to which the ratios listed above all belong.

Looking first at the examples in Realm 1, we see from the tables that the hiker is walking at a steady rate: in twice the time, twice the distance is covered, and so on. In the second example, the labourers are all working at the same rate as each other: twice their number lay twice the volume of concrete, and so on. In the third example, the path rises at a steady rate. So, in the mathematical model common to all these examples, we have a set of ratios which are equivalent in that they all represent *the same rate*.

We now want a mathematical test which we can use to find whether any two given ratios are equivalent in this way. Keeping for the moment to the same equivalence class, we can see that $5:2 \cong k \times 5:k \times 2$ whatever natural number k stands for. Multiply the numbers in pairs as indicated below

$$5:2 \cong k \times 5:k \times 2$$

and we get from the outer pair $5 \times k \times 2$
and from the inner pair $2 \times k \times 5$

Each product contains a 5 from one ratio, a 2 from the other ratio and a k from one term or other of the second ratio. So it is not fortuitous that the results are equal – they must be so.

It is easy, with the help of a little algebra, to turn this result into a general test of equivalence* between ratios.

$$a:b \cong a':b' \qquad \text{if and only if} \qquad ab' = ba'$$

Practise this test of equivalence for yourself with ratios taken at random from the set $5:2$, $10:4$, $15:6$, $20:8$, etc.; and experiment a little with other sets of equivalent ratios, constructing and testing them in the same way. (To construct an equivalence class, write any ratio at random, say $12:4$; and either multiply both terms by the same natural number, giving $24:8$, $120:40$, etc., or, if it can be done, divide both terms by the same natural number, giving $6:2$, etc.)

A set of equivalent ratios is called a *proportion*. We say, for example, that $5:2$ and $30:12$ are *in the same proportion*. A rational number is the characteristic property of a proportion, and is represented by any of its ratios.

Adding and multiplying rational numbers

Adding. Suppose a man is walking up an escalator at a rate given by this table.

Steps	5	10	15	20 ...
Seconds	2	4	6	8 ...

* This is the only kind of equivalence between ratios which is in general use, so we do not need to specify it further.

At the same time the escalator is moving up at this rate.

Steps	3	6	9	12	15	18 ...
Seconds	1	2	3	4	5	6 ...

His total rate of ascent is given by this table. We can, of course, only add the numbers of steps which correspond to the same number of seconds.

Steps	5 + 6	10 + 12	15 + 18 ...
Seconds	2	4	6 ...

This suggests that addition of rational numbers might usefully (in terms of being a sensible model) be defined as the *sum* of two ratios, thus:

$$5:2 \oplus 6:2 = (5 + 6):2$$
$$10:4 \oplus 12:4 = (10 + 12):4, \text{ etc.}$$

and in general $\qquad a:b \oplus a':b = (a + a'):b$

If the ratios have not the same second term as each other, we must replace them by equivalent ratios which have.

$$a:b \oplus a':b'$$
$$= ab':bb' \oplus a'b:bb' = (ab' + a'b):bb'$$

Multiplying. Consider now these two tables, which show respectively the correspondences between kilometres travelled and litres of petrol consumed by a car, and between petrol consumed and time.

Distance in kilometres	9	18	36	54	72	90	108 ...
Fuel in litres	1	2	4	6	8	10	12

Fuel in litres	3	6	9	12 ...
Time in hours	1	2	3	4

We can combine these tables to give correspondences between distance travelled and time. In this case, we have to pick pairs for which the fuel figure is the same.

Distance in kilometres	54	108 ...
Time in hours	2	4 ...

This suggests that multiplication of rational numbers might reasonably be made to correspond to the *product* of the two rates. In this case,

$$54:6 \otimes 6:2 = 54:2$$
$$108:12 \otimes 12:4 = 108:4, \text{ etc.}$$
and in general $\qquad a:b \otimes b:c = a:c$

If the second term of the first ratio and the first term of the second ratio are not equal, we must replace these ratios by equivalent ones for which this is so.

$$a:b \otimes a':b' = aa':ba' \otimes ba':bb'$$
$$= aa':bb'$$

At this stage, it would be a useful exercise for the reader to verify that the interchangeability applies, using some of the numerical examples above. For example, that the results of

$$5:2 \oplus 6:2 \quad \text{and} \quad 15:6 \oplus 18:6$$

are equivalent ratios (equal rational numbers).

Negative ratios

In the escalator example, there is no reason why the man and the escalator have to be moving in the same direction. If the man is walking down, while the escalator is moving up, both at the same speed as before, we now require positive and negative integers for our mathematical model. Up is the direction more usually chosen to be represented by positive integers, but there is no hard and fast rule about this.

Man	Steps	(-5)	(-10)	(-15)	(-20) ...		
	Seconds	$(+2)$	$(+4)$	$(+6)$	$(+8)$ ...		
Escalator	Steps	$(+3)$	$(+6)$	$(+9)$	$(+12)$	$(+15)$	$(+18)$ ...
	Seconds	$(+1)$	$(+2)$	$(+3)$	$(+4)$	$(+5)$	$(+6)$ ...

The two proportions are represented by $(-5):(+2)$ and $(+6):(+2)$, and their sum, representing the man's total rate of ascent, is $(-5):(+2) \oplus (+6):(+2)$, which is equal to $(+1):(+2)$.

In the motoring example, distance might well be measured from some point on the route, in which case distances before and after passing it would be represented by positive and negative integers. And, in general, since things decrease as well as increase, if our ratios are to act as general-purpose models for all kinds of rate, they will have to be ordered pairs of integers, not of natural numbers. Looking back now to the first escalator example, we can see quite a strong case for using integers there also, rather than natural numbers, since the numbers were being used to represent reversible events rather than countable objects. However, since natural numbers and positive integers (or rather non-negative integers) are isomorphic systems, we usually save trouble by working in the simpler one.

Are the rationals a number system? *

Our provisional definitions of $\oplus$ and $\otimes$ on the rational numbers were based on their suitability for mathematical models for sets of equivalent rates. Final acceptance depends on their satisfying the five requirements for a number system, which they can easily be seen to do. For simplicity of adding, we shall choose ratios which have the same second term. Let a, b, c, d, m, n, stand for any integers.

ADDITION IS COMMUTATIVE

$$a{:}d \oplus b{:}d = (a + b){:}d = (b + a){:}d = b{:}d \oplus a{:}d$$

ADDITION IS ASSOCIATIVE

$$(a{:}d \oplus b{:}d) \oplus c{:}d = (a + b){:}d \oplus c{:}d = [(a + b) + c]{:}d$$
$$= [a + (b + c)]{:}d = a{:}d \oplus (b + c){:}d = a{:}d \oplus (b{:}d \oplus c{:}d)$$

Proofs that $\otimes$ is commutative and associative are similar. Readers may care to write their own, as an exercise.

MULTIPLICATION IS DISTRIBUTIVE OVER ADDITION

The choice of different letters for the first ratio is simply to make it look different, and thus show the pattern more clearly.

$$m{:}n \oplus [a{:}d \oplus b{:}d] = m{:}n \otimes (a + b){:}d$$
$$= m(a + b){:}nd$$
$$= (ma + mb){:}nd$$
$$= ma{:}nd \oplus mb{:}nd$$
$$= m{:}n \otimes a{:}d \oplus m{:}n \otimes b{:}d$$

* May be omitted without loss of continuity.

As with fractional numbers and natural numbers, so also with rational numbers and integers: the new system has the five properties because (but not only because) the earlier number system has them.

Rational numbers in fractional notation

Though their origins are so different, rational numbers derived from positive integers are isomorphic with fractional numbers. Compare, for example,

$$(+5):(+2) \oplus (+6):(+2) = (+11):(+2) \text{ and } \tfrac{5}{2} \oplus \tfrac{6}{2} = \tfrac{11}{2}$$

and the general case

$$a:d \oplus b:d = (a + b):d \text{ and } \frac{a}{d} \oplus \frac{b}{d} = \frac{a + b}{d}$$

A similar isomorphism exists for $\otimes$. Moreover, the conditions for equivalence of ratios, and for equivalence of the corresponding fractions, are almost the same.

$$\left. \begin{array}{c} a:b \cong a':b' \\[2mm] \dfrac{a}{b} \cong \dfrac{a'}{b'} \end{array} \right\} \text{ iff } ab' = a'b$$

The only difference is that in the ratios, a, b, a', b', represent positive integers, while in the fractions they represent natural numbers. So for calculations, we can take advantage of the simpler fractional notation.

This advantage can, moreover, be extended to ratios involving negative integers, provided that we now use the appropriate operations.

Instead of $(+5):(+2) \oplus (-6):(+2) = (-1):(+2)$

we can write $\dfrac{(+5)}{(+2)} \oplus \dfrac{(-6)}{(+2)} = \dfrac{(-1)}{(+2)}$

noticing that the (-1) on the right-hand side was obtained by the operation for adding *integers*.

The above is usually written, loosely but conveniently,

$$\frac{5}{2} - \frac{6}{2} = -\frac{1}{2}$$

Here, $-\dfrac{6}{2}$ could mean either $\dfrac{(-6)}{(+2)}$ or $\dfrac{(+6)}{(-2)}$

However, when using fractional notation for rational numbers, it is usual to keep to positive denominators, replacing, for example, $\dfrac{(+5)}{(-2)}$ by the equivalent fraction $\dfrac{(-5)}{(+2)}$, which is then written $\dfrac{-5}{2}$, or even $-\dfrac{5}{2}$.

One could devote considerable space to a consideration of this apparently slipshod notation, from two points of view. The first question is why it works at all, and we have already seen, in general terms, that the answer lies in the isomorphisms connecting natural numbers, fractional numbers and rational numbers – though there is still much detail to be filled in. The second question is why an inaccurate notation can actually be better for certain purposes than an accurate one. The convenience of statements like $\frac{5}{2} - \frac{6}{2} = -\frac{1}{2}$ seems to derive largely from their looseness – that they allow us to ignore important facts, such as what kind of number we are talking about, and thereby to simplify our calculations. This kind of ignoring may be due to ignorance or it may be based on knowledge of what may validly be disregarded in certain circumstances. In the former case we get the right answer more by luck than by judgement.

Ratios written as decimal fractions

When first constructing a model, the original ratio notation has the advantage of showing more clearly the origin of the pairs of numbers in sets of corresponding events. One can then change over to fractional notation, which is more convenient for calculation. The latter also leads directly to another way of representing a rational number, which forms an essential preliminary to the next number system.

If we write a ratio, say, $(+5):(+2)$, in fractional notation as $\dfrac{(+5)}{(+2)}$, and replace this in turn by the fractional number $\dfrac{5}{2}$, we can add a fourth term to this chain of correspondences, namely, 2·5, the same (fractional) number written in decimal notation. Can we use 2·5 as an alternative notation for the original rational number $(+5):(+2)$ also? And, in general, is decimal notation a possible one for rational numbers including calculations?

The fraction 2·5 is equal to the fraction $\frac{25}{10}$. All that is different is the notation. Let us now agree that $(+2\cdot5)$ means the same as $\frac{(+25)}{(+10)}$, a ratio written in fractional notation. Let us also agree that $(-2\cdot5)$ means the same as both $\frac{(-25)}{(+10)}$ and $\frac{(+25)}{(-10)}$, these ratios being equivalent. Since calculations with fractional numbers can be done either* in common or decimal notation, we can now say the same for rational numbers, provided that whenever necessary we replace the operations for natural numbers by those for integers. For example,

$$(-2\cdot5) \oplus (+1\cdot2) = (-1\cdot3)$$
$$(-2\cdot5) \otimes (+1\cdot2) = (-3\cdot00)$$

As a check, let us work the second of these products of rational numbers in (common) fractional notation. We had better keep to the accurate form of this.

$$\frac{(-25)}{(+10)} \otimes \frac{(+12)}{(+10)} = \frac{(-25) \otimes (+12)}{(+10) \otimes (+10)} = \frac{(-300)}{(+100)}$$

Notice that the $\otimes$ on the left represents the multiplication of rational numbers, while that in the centre represents that of integers. We should really use different signs, but suitable ones are not available in printer's type. Notice also that the last is equivalent to $\frac{(-3)}{(+1)}$, but that this is not quite the same as (-3). A ratio has two terms, so if we find an integer apparently doing duty for a rational number, this means that the second term is $(+1)$ which has got lost, probably by the use of the 'loose but convenient' notation referred to earlier.

Decimal notation is a convenient and widely used notation for rational numbers. In this case the second term is present implicitly. For example,

* We often use a mixture: for example, $\dfrac{11\cdot7 \times 2\cdot4}{8\cdot25}$

0·75 represents the same fractional number as $\frac{75}{100}$. It also represents the same rational number as 75:100.

Real numbers

An Egyptian mural three thousand years old shows surveyors carrying rope in which are 12 equidistant knots. When such a rope is pulled into a triangle with sides of length 3, 4, 5 units, the triangle is right-angled. This result uses (in reverse) the famous theorem known as Pythagoras'

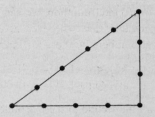

theorem: that in any right-angled triangle, if we draw squares on the three sides, the area of the square on the hypotenuse (the side opposite the right angle) is equal to the sum of the areas of the squares on the other two sides. In the case of the 3, 4, 5 triangle shown above, the areas can be found by counting the unit squares. We could also find them by

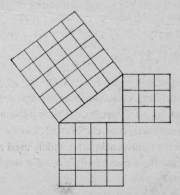

calculating, the areas being respectively 3 × 3, 4 × 4, 5 × 5 square units: that is, 9, 16, 25 square units.

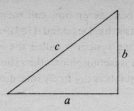

In the general case, if the sides are of lengths, a, b, c, units respectively, the theorem states that

$$a^2 + b^2 = c^2$$

where $a^2 = a \times a$ and likewise for b and c.

(If a square has sides of length a units, then its area is $a \times a$ square units, so $a \times a$, or a^2, is usually read as 'a squared' rather than 'a to the power 2'.)

This interesting but innocuous looking theorem had, and has, some surprising consequences. Suppose that we have a right-angled triangle

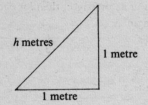

whose two shorter sides both have lengths 1 metre. Then we should be able to calculate the length of the hypotenuse by finding a number h such that

$$h^2 = 1^2 + 1^2$$

that is, $$h^2 = 2$$

since $$1^2 = 1$$

But in the rational number system, we cannot find a number which when squared (multiplied by itself) gives 2. What is more, it is not hard to prove that no such rational number can exist, though we shall not digress to do so here. The Pythagoreans, who made their mathematics into a semi-religious cult, knew of this proof; and the existence of this and other non-rational numbers was so upsetting to their established ways of thinking that they called them 'unspeakable numbers', and swore their members to secrecy. (This example of the difficulty of restructuring one's schemas has been mentioned already.)

These irrational numbers, as we now call them, have been put on a sound mathematical basis by Dedekind (1831–1916) and by Cantor (1845–1918). The resulting system is called the *real number* system. It contains a sub-set which is isomorphic to the rational number system, so rational and irrational numbers are freely intermixed, as with the earlier number systems.

Nests of intervals

Of the two approaches, Cantor's is probably easier for the general reader to understand. We shall assume a knowledge * of multiplication of rational numbers in decimal notation, and illustrate the method by trying to find the number whose square is 2. (This number is called the *square root* of 2, written $\sqrt{2}$.)

If we mark off a line in units of length, then for every rational number there corresponds a length on the line. If all lengths are measured from the same zero point, then for each rational number there is a point on the line.

Here are the points which correspond to 0·7, 1·0, 1·5, 2·16. (In this section we shall for simplicity keep to positive rationals.)

We have found that there is a length, and so a point, for which there is no corresponding rational number; the problem is to invent a number of some sort which does

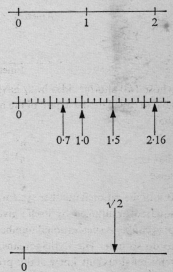

---

* Reminder: one way is to omit the decimal point, multiply, and replace the decimal point by calculating an approximate answer in integers; another way is to use a calculator! Note that calculators can work in natural numbers, integers, and rational numbers in decimal notation, but not in real numbers.

correspond to this point. Calling it a number at this stage is simply a working hypothesis, and writing it as $\sqrt{2}$ is simply a short way of stating the condition it has to satisfy, that

$$\sqrt{2} \times \sqrt{2} = 2$$

Clearly 1 is too small, for $1 \times 1 = 1$
and   2 is too big, for $2 \times 2 = 4$.
So we know that the
point corresponding to $\sqrt{2}$
lies between those
corresponding to 1 and 2
on the line.

Let us find a smaller interval in which $\sqrt{2}$ also lies.

$$1\cdot1 \times 1\cdot1 = 1\cdot21 \quad \text{Too small}$$
$$1\cdot2 \times 1\cdot2 = 1\cdot44 \quad \text{Too small}$$
$$1\cdot3 \times 1\cdot3 = 1\cdot69 \quad \text{Too small}$$
$$1\cdot4 \times 1\cdot4 = 1\cdot96 \quad \text{Too small}$$
$$1\cdot5 \times 1\cdot5 = 2\cdot25 \quad \text{Too big}$$

So the point corresponding
to $\sqrt{2}$ lies between those
corresponding to 1·4 and
1·5.

Now to find a smaller interval still.

$$1\cdot41 \times 1\cdot41 = 1\cdot9881 \quad \text{Too small}$$
$$1\cdot42 \times 1\cdot42 = 2\cdot0164 \quad \text{Too big}$$

So the point corresponding
to $\sqrt{2}$ lies between those
corresponding to 1·41 and
1·42.

This is now getting too small to draw. Nevertheless, we can calculate more pairs of rational numbers whose corresponding points enclose ever diminishing intervals of the line. The next three in the sequence are

1·414	and	1·415
1·4142	and	1·4143
1·41421	and	1·41422

Given time, patience and adequate calculating facilities, we can continue this narrowing-down process as long as we like.

This sequence of intervals on the line has two interesting properties:

(i) Each interval lies within the one before it.

(ii) However small a number we care to name, we can find an interval such that the size of this interval *and of all its successors* is smaller than this number.

For example, the size of the last-named interval, and of all that follow it, is less than 0·0001.

Any such sequence of intervals is called a *nest* of intervals, and it is easy to prove that there is only one point which is in *every* interval of the nest. For if there were two such points, they would have to be some distance apart, however small, and we could then find an interval whose size was smaller than this distance. This would put one of the points outside the interval.

It follows that, in this example, the infinite sequence of pairs

1	and	2
1·4	and	1·5
1·41	and	1·42
1·414	and	1·415
1·4142	and	1·4143
etc.		etc.

defines uniquely the point corresponding to $\sqrt{2}$. This is what we set out to do. The number thus (and rather lengthily) defined is called a *real number*.

What we have done is show that the number which – assuming in advance its existence – we call $\sqrt{2}$ can be represented in terms of rational numbers, just as earlier we showed that rational numbers can be represented in terms of integers and integers in terms of natural numbers. To justify its being called a number, ways of adding and multiplying this and other numbers similarly defined, which result in a number system, must now be found.

Adding and multiplying real numbers

To get some suggestions for suitable methods of adding and multiplying real numbers, we may recall that they have originated from a situation in which the measure of the length of a line turned out to be not expressible as a rational number of units. What the sequence of intervals tells us is that this length (in the present example, $\sqrt{2}$ units) lies between 1·4 and 1·5 units, between 1·41 and 1·42 units, etc.

The diagram represents
the first of these intervals.

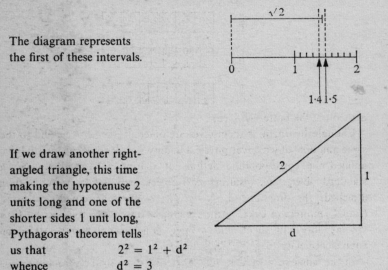

If we draw another right-angled triangle, this time making the hypotenuse 2 units long and one of the shorter sides 1 unit long, Pythagoras' theorem tells us that $2^2 = 1^2 + d^2$
whence $d^2 = 3$

A sequence of calculations like those for $\sqrt{2}$ tells us that

$\sqrt{3}$ is between 1 and 2
between 1·7 and 1·8
between 1·73 and 1·74
between 1·732 and 1·733
etc.

Adding these two numbers, $\sqrt{2}$ and $\sqrt{3}$, should correspond to putting the two line segments* end to end (in a straight line) and finding out how long it is.

---

* A straight line is considered to extend indefinitely in both directions, so if we mean that part of a line which lies between two points on it we call it a line segment.

Since $\sqrt{2}$ is between	and $\sqrt{3}$ is between	$\sqrt{2} \oplus \sqrt{3}$ is between
1 and 2	1 and 2	2 and 4
1·4 and 1·5	1·7 and 1·8	3·1 and 3·3
1·41 and 1·42	1·73 and 1·74	3·14 and 3·16
1·414 and 1·415	1·732 and 1·733	3·146 and 3·148
etc.	etc.	etc.

Without giving a formal proof, it is easy to see that this way of adding gives another nest of intervals, thereby defining another unique real number. This addition is commutative and associative, since the intervals which define ($\sqrt{2} \oplus \sqrt{3}$) result from adding rational numbers.

If we draw a rectangle with sides 1·4 units and 1·7 units, its area will be $1·4 \times 1·7 = 2·38$ square units. Similarly, if we draw a rectangle with sides 1·5 and 1·8 units, its area will be $1·5 \times 1·8 = 2·70$ square units.

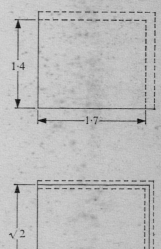

If we now draw another rectangle with sides $\sqrt{2}$ and $\sqrt{3}$ units, its area will be between the two areas which we have just calculated. So it seems reasonable to insist that however we define* $\sqrt{2} \oplus \sqrt{3}$, this real number must be between 2·38 and 2·70.

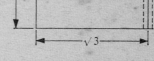

But this argument is equally valid if we replace these two rectangles by others with dimensions $1·41 \times 1·73$ and $1·42 \times 1·74$ units, or $1·414 \times 1·732$ and $1·415 \times 1·733$ units.

* Using $\otimes$ for multiplication of reals and $\times$ for that of rationals.

Since $\sqrt{2}$ is between	and $\sqrt{3}$ is between	$(\sqrt{2} \otimes \sqrt{3})$ is between
1 and 2	1 and 2	1 and 4
1·4 and 1·5	1·7 and 1·8	2·38 and 2·70
1·41 and 1·42	1·73 and 1·74	2·4393 and 2·4708
1·414 and 1·415	1·732 and 1·733	2·449048 and
etc.	etc.	2·452195 etc.

It seems clear that we have another nest of intervals in the right-hand column, thereby defining another real number. It is also clear that multiplication defined in this way is commutative and associative. That it is distributive over addition can be shown by a combination of the two methods above. This is not especially difficult, but there is more detail to handle and we shall not attempt this here. Readers may, however, care to check for themselves that these methods apply equally well to negative real numbers, defined by nests of intervals between negative rationals.

Mixed real and rational numbers

A nest of intervals is the most complicated way of representing a number which we have yet encountered. We can simplify it considerably by the following notation, called the 'infinite decimal' notation.

1·4 . . .   means 1·4 'and a bit', that is, between 1·4 and 1.5
1·41 . . .   means between 1·41 and 1·42
1·414 . . . means between 1·414 and 1·415

This means that reals and rationals can be mixed for purposes of calculation, replacing the real numbers by their rational approximations: for example, replacing $\sqrt{2}$ (= 1·414 . . .) by 1·414 without the continuation dots. But we have to be careful in assessing the degree of accuracy of our answer. It would be tempting to write

$$\sqrt{2} \otimes \sqrt{3} = (1 \cdot 414 \ldots) \times (1 \cdot 732 \ldots)$$
$$= 2 \cdot 449048 \ldots$$

but this would mean that the result was between 2·449048 and 2·449049, which we do not know. Looking back to the nest of intervals defining

$\sqrt{2} \otimes \sqrt{3}$, all we can say for sure is that the result is between 2·449048 and 2·452195. For a similar reason we cannot even write

$$\sqrt{2} \otimes \sqrt{3} = 2·449\ldots$$

Readers should verify for themselves that in fact all we can say from the above in the infinite decimal notation is

$$\sqrt{2} \otimes \sqrt{3} = 2·4\ldots$$

They should also explain to their own satisfaction the seeming paradox:

$$1·414 \times 1·732 = 2·449048 \text{ (rationals)}$$
but $\qquad (1·414\ldots) \otimes (1·732\ldots) = 2·4\ldots \text{(reals)}$

Irrational numbers

Some square roots can be calculated exactly: for example, $\sqrt{4}$, which is equal to 2. To avoid having gaps in the real numbers, we can represent this by the infinite decimal 2·000 . . ., or by this nest of intervals:

$\sqrt{4}$ is between 2 and 3
2·0 and 2·1
2·00 and 2·01
2·000 and 2·001
etc.

In this nest, the real number $\sqrt{4}$ coincides with the lower number in every interval, so 'between' must now enlarge its meaning to include the end points of the interval.

Real numbers which have no corresponding rational number are called *irrational numbers*. Though square roots were our original source of irrational numbers, there are many others. Indeed, any infinite decimal which does not repeat itself represents an irrational number, and Cantor has also shown that there are many more irrational numbers than rational.

Looking back

In the last two chapters we have covered ground which would normally be spread over about ten years at school, and which took over twenty-three centuries in the history of mathematics. Looking back, we can see the whole progress as an outstanding example of simultaneous assimilation and expansion, at each stage where a new number system is introduced.

Assimilation	*Expansion*
Each new system continues to use the Hindu-Arabic numerals of the original natural number system	together with new signs such as the fraction bar, decimal point, positive/negative signs.
Each new system continues to use the operations of addition and multiplication	but gives them new meanings.
The new methods for adding and multiplying continue to use the results memorized for the natural number system	adapted by simple additional rules to the new system.
The five properties of a number system are retained	with new operations.

In each new system there are sub-sets which are isomorphic with the earlier systems. This allows us to move freely from one number system to another and also to mix the systems, provided that each is operated according to its own methods. The overall result is a conceptual system of enormous power and flexibility.

CHAPTER 12

Algebra and Problem-solving

In the last few chapters we have studied the development of some powerful conceptual tools – number systems and systems of numerals whereby these numbers can be easily recorded and manipulated. In this chapter we shall look at the beginnings of algebra, which generalizes certain ideas of the number schema in a new direction, and we shall examine how algebraic statements can also serve as models for problems in the physical world, and how algebraic manipulations can be used to find solutions of these problems.

The idea of a variable

The idea of a set and that of a variable are two of the most basic in mathematics. The first of these has already been made explicit and put to work. The second has been used implicitly in a few places where we could not do without it; it is now time to bring this idea also from the intuitive to the reflective level. The idea of a variable is in fact a key concept in algebra – although many elementary texts do not explain or even mention it.

Often in everyday life we refer to an object in a way which indicates what set it belongs to but not which particular element of the set it is. For example, 'a car' is a shorter way of saying 'an unspecified element of {motor cars}'. 'A handkerchief' means 'an unspecified element of {handkerchieves}'. We talk in this way for various reasons. One is when it does not matter which: if we want to blow our nose, any handkerchief will do. Another is when we want to make statements which are true for each and every element of the set, and do not depend on which element we refer to. Examples: a car must have a licence; a car must stop at red traffic lights; a car needs petrol and oil.

A third reason for not specifying which element of a set we are referring to is when we do not know: for example, my wallet has been stolen by a pickpocket. In cases of this last kind, we often wish to know. Finding out often presents a problem, however, and it is an advantage if there already exist well-developed methods for solving problems of the various kinds which we encounter.

In mathematics, an unspecified element of a given set is called a *variable*. Variables are often referred to in the same way as in everyday speech: for example, 'a number', 'a circle', 'a set'. But for many purposes other symbols are more convenient. In geometry, it helps our thinking to draw a particular circle (or square, or triangle), but to ignore any properties which depend on which particular one we happen to have drawn. When our variable is a number, the likelihood of confusion becomes greater, so we represent the variable by a letter. We can now make statements like:

If a, b, are any numbers, then

$$a + b = b + a$$

which the reader will recognize as the first of the characteristic properties of a number system. So the idea of a variable was already being used implicitly on page 161, where it enabled us to make briefly and clearly certain statements which are true for any numbers. Implicit also is the assumption that 'a' stands for the same number on both sides of the equation, though we have not said which number it is, and likewise for 'b'. It is not, however, implied that 'a' and 'b' represent different numbers; clearly this statement is true if both stand for the same number, though it then becomes trivial. The elements which a variable may stand for are called *values* of the variable, so we could express the above by saying that a must have the same value on both sides of the equation, and that a and b may have different values from each other, or the same.

Letters are freely used as symbols for variables of all kinds. Using $\cup$ as a symbol for the operation of uniting two sets (see Chapter 8), we may write:

If A, B, are any sets, then

$$A \cup B = B \cup A$$

It is therefore necessary to know what kinds of variable we are talking about.

Algebra

Algebra is concerned with statements involving variables of any kind. There is, for example, an algebra of sets, to which the above statement belongs. There is an algebra of functions, which we shall look at in Chapter 13. For the rest of this chapter, however, we shall be concerned with the algebra of numbers, so all the variables referred to will be numerical variables – unspecified numbers. Unless stated otherwise, they can be any kind of number: natural number, integer, fractional, rational or real number. This is the first kind of algebra to have been developed, and it is still probably the most widely used, so by itself 'algebra' usually means the algebra of numerical variables.

First, a few details of notation. For the most part letters representing variables and numerals representing specific numbers are freely intermixed using the same operational and relational signs for both. For example,

we write $\qquad 7 + 7 + 7 = 3 \times 7$
and $\qquad a + a + a = 3 \times a$
Also $\qquad 7 \times (4 + a) = 28 + (7 \times a)$

Notice in the last equation, which uses the distributive property, that we can complete the calculation 7×4 but not the calculation $7 \times a$, since we do not know what number a is. For the same reason, it is usually the case in algebra that we cannot carry our calculations to completion, as in arithmetic we usually can. This may be why the following two notations are generally used, by which these uncompleted calculations can be written more shortly.

Where sums and products are mixed, it is agreed that products are calculated before sums. This allows us to leave out parentheses in many cases.

For example, $a + b \times c$ always means $a + (b \times c)$
$\qquad\qquad\qquad\qquad$ not $(a + b) \times c$

Secondly, for products involving variables the multiplication sign may also be omitted.

For example, $b \times c$ may be written $\qquad bc$
$\qquad\quad 2 \times b$ may be written $\qquad 2b$
$\quad$ (But 2×7 is, of course, never written $\quad 27$
$\qquad\qquad$ since this would mean $\qquad$ 2 tens and 7 units.)

So $a + (b \times c)$ is usually written as $a + bc$ which is certainly shorter. Nevertheless, this condensed notation can be misleading, since it is inconsistent with place value notation in arithmetic. 'bc' does not mean 'b tens and c units' – both b and c could be numbers each of many digits. Also 'bc' and 'cb' both represent the same number (since multiplication is commutative), while '27' and '72' do not. Care is specially needed when the two notations are mixed.

For example, $7(4 + a)$
is equal to $28 + 7a$
and not $74 + 7a$

It is a help initially, until the meaning of the notation is well established, to read, for example,

bc as b times c
$7(4 + a)$ as   7 times (4 plus a)

The advantages of this notation outweigh its disadvantages, particularly when complicated products are involved. And it does help us to think of a product as a single number, and thereby to remember that

$a + bc$ means $a + (bc)$
    and not $(a + b)c$

It is not, however, compulsory, and we are free to use it or not, according to whether it helps or hinders in a particular context.

Index notation. Products of a number by itself are written still more briefly by the use of index notation.

$a \times a$ is written not aa but a^2
$a \times a \times a$ is written a^3
$a \times a \times a \times a$ is written a^4, and so on.

The last of these is read as 'a to the fourth power', or 'a to the fourth' for short, with similar meanings for a^5 ('a to the fifth power'), a^6, and so on. But a^2 is usually read as 'a squared', since this is the area of a square whose side is of length a units. For a similar reason, a^3 is read as 'a cubed'. However, it would be perfectly correct, and more consistent, to read these as 'a to the second', 'a to the third'. In, say, a^6, a is called the *base*, and 6 is called the *index*.

A method for simplifying products of various powers of the same base is easily developed, from first principles.

Example:
$$a^2 \times a^3 = a \times a \quad \times \quad a \times a \times a$$
$$= a \times a \times a \times a \times a$$
$$= a^5$$
Similarly:
$$a^5 \times a^7 = a \times a \times a \times a \times a \quad \times$$
$$a \times a \times a \times a \times a \times a \times a$$
$$= a^{12}$$
And in general:
$$a^m \times a^n = a^{m+n}$$

Notice that we have now introduced two more variables, m and n. The first variable, a, enabled us to write or talk about *specified* powers of *any* base. These two new ones, m and n, allow us to talk about *any* powers of *any* base.

When using variables, we have to know what set they belong to. Can m and n belong to *any* number system? At present, no. The original definition of index notation has meaning only if the indices are natural numbers, excluding zero. But in Chapter 3, page 57, it was mentioned (as an example of mathematical generalization) that meanings could be found for a^0, a^{-1}, a^{-2}, etc. and also for $a^{\frac{1}{2}}$, $a^{\frac{1}{3}}$, etc. These are arrived at as follows.

First, we extend the original method to division.
Example:

$$a^5 \div a^3 = \frac{a \times a \times a \times a \times a}{a \times a \times a}$$

$$= \frac{a \times a}{1} \qquad \text{(by earlier work on fractions)}$$

$$= a^2$$

$$= a^{5-3}$$

By a similar method, we can see that, for example,
$$a^7 \div a^2 = a^5 = a^{7-2}$$
$$a^8 \div a^6 = a^2 = a^{8-6}$$
and in general $a^m \div a^n = a^{m-n}$

At present a^{m-n} has meaning only if $m > n$ (m is greater than n), since otherwise the index is negative or zero. What happens if we remove this restriction?

Suppose that m and n both have the value 3.

This gives $\qquad\qquad a^3 \div a^3 = a^0$ by the general method.

From first principles $a^3 \div a^3 = \dfrac{a \times a \times a}{a \times a \times a}$

$$= 1$$

This suggests that a reasonable meaning for a^0 is 1 – reasonable in the sense that it is consistent so far with our existing schema for indices. Let us test it further in some other examples.

$$a^2 \times a^0 = a^{2+0} = a^2$$

This is consistent with $\qquad a^2 \times 1 \ = a^2$

$$a^2 \div a^0 = a^{2-0} = a^2$$

which is consistent with $\qquad a^2 \div 1 \ = a^2$

So we shall settle for 1 as the meaning of a^0.

Now what about $a^0 \div a^2$?

The existing method for division gives $a^0 \div a^2 = a^{0-2} = a^{-2}$

Replacing a^0 by 1, $\qquad\qquad\qquad 1 \div a^2 = \dfrac{1}{a^2}$

So we can retain our existing method for division if we agree that a^{-2} now means $\dfrac{1}{a^2}$ (up till now, a^{-2} has had no meaning). This works also for multiplication. For example,

$$a^5 \times a^{-2} = a^{5+(-2)} = a^{5-2} = a^3$$

is consistent with

$$a^5 \times \frac{1}{a^2} = \frac{a^5}{a^2} = a^3$$

Similar arguments indicate that consistency with existing methods will be obtained by giving

$$a^{-1} \text{ the meaning } \frac{1}{a}$$

$$a^{-3} \text{ the meaning } \frac{1}{a^3}$$

and in general a^{-m} the meaning $\dfrac{1}{a^m}$

What we have done here is not *quite* the same as assimilating new ideas to an existing schema. Rather, from various possibilities ($a^0 = 0$ is one which is often favoured by younger students) we have chosen new ideas on the basis of their assimilability. Such expansion as was required of the schema did not involve changing its most useful features, the quick methods for multiplying and dividing powers of the same base, but only abandoning restrictions that the indices had to be non-zero natural numbers. This is a good way to progress!

To find a meaning for fractional powers such as $a^{\frac{1}{2}}$, we must first know another general method for working with indices. Here is a particular case.

$(a^2)^3$ meaning the third power of (a to the second power)

$$= a^2 \times a^2 \times a^2$$
$$= a^{2+2+2}$$
$$= a^6$$
$$= a^{2 \times 3}$$

This method clearly does not depend on our particular choice of the numbers 2 and 3, so we may replace them by variable natural numbers.

$$(a^m)^n = a^{mn}$$

Now suppose we remove the restriction that m has to be a natural number and give it the value $\frac{1}{2}$. If we also take $n = 2$, then to preserve this method under the new condition, we must give $a^{\frac{1}{2}}$ whatever meaning will make the following statement true.

$$(a^{\frac{1}{2}})^2 = a^{\frac{1}{2} \times 2} = a^1 = a$$

In words, we must take $a^{\frac{1}{2}}$ to mean that number whose square (second power) is equal to a. We call this the *square root of a* (see Chapter 11, page 205, where the term was first introduced). So by agreeing that $a^{\frac{1}{2}} = \sqrt{a}$, we have made index notation meaningful for the fractional number $\frac{1}{2}$. We must, however, restrict both a and its roots to positive real numbers, to avoid complications which we do not wish to pursue here.

By a similar argument, we can generalize index notation to fractional indices of the form $\frac{3}{4}$, $\frac{1}{4}$, etc., by letting $a^{1/n}$ mean *the nth root of a*, that is the number x such that $x^n = a$. This is sometimes written $\sqrt[n]{a}$. To find a suitable meaning for $a^{m/n}$ we can write this as either $(a^{1/n})^m$ or as

$(a^m)^{1/n}$. The first means the mth power of $a^{1/n}$, the second means the nth root of a^m. Without embarking on a general proof, let us check in a particular case that these are the same number. This will also serve to illustrate the meaning of a fractional power, by a particular example. To avoid having to work with irrational numbers, we will choose

> for a the value 81
> for m the value 3
> for n the value 4

Replacing the variables a, m, n, by these values,
the first expression $(a^{1/n})^m$ becomes

$$(81^{\frac{1}{4}})^3 = (3)^3 \qquad \text{since } 3^4 = 81$$
$$= 27$$

The second expression $(a^m)^{1/n}$ becomes

$$(81^3)^{\frac{1}{4}} = (531441)^{\frac{1}{4}}$$
$$= 27 \qquad \text{since } 27^4 = 531441$$

The reader may like to verify other particular cases as an exercise. Convenient values are $a = 125$, $m = 2$, $n = 3$; and $a = 32$, $m = 3$, $n = 5$. (Note: $125 = 5^3$ and $32 = 2^5$.)

Calculations in algebra

To work with an arithmetical model, we have to learn to do calculations with specified numbers: for example, $17 + 44$, 51×65, $8 \times 3 \cdot 142$. To work with an algebraic model, we have to become equally competent at calculations with variables. These are for the most part easier, since all we have to do to multiply the unspecified numbers a and b is to write them as ab. To add them, we write $a + b$, and if we like enclose them in parentheses $(a + b)$ to show that we are now thinking of the single number which results from adding a and b.

All algebraic calculations are built up from the five characteristic properties of a number system. Since these properties are true for natural numbers, integers, fractional, rational and real numbers, it follows that these algebraic methods will apply equally well to all. For easy reference, here are the five properties.

$$a + b = b + a$$ addition is commutative
$$(a + b) + c = a + (b + c)$$ addition is associative
$$ab = ba$$ multiplication is commutative
$$(ab)c = a(bc)$$ multiplication is associative
$$a(b + c) = ab + ac$$ multiplication is distributive over addition

The distributive property, used backwards, enables us to simplify expressions like $5a + 3a$.

$$5a + 3a = (5 + 3)a = 8a$$

Notice that we have here used the distributive property in the form $ba + ca = (b + c)a$. This is true because $(b + c)a = a(b + c)$, that is, because multiplication is commutative.

Another example: to simplify $4a + 3b + 7a + 2b$

First we use the commutative property of addition to rearrange the order. This gives

$$4a + 7a + 3b + 2b$$

Then the distributive property in reverse, as before.

$$= (4 + 7)a + (3 + 2)b$$
$$= \quad 11a \quad + \quad 5b$$

Implicit also is the associative property of addition, that it makes no difference which terms we add together first. Sometimes we use the distributive property first forwards, then backwards.

Example: to simplify $3(2a + 5b) + 6(4a + 7b)$
This is equal to $3 \times 2a + 3 \times 5b + 6 \times 4a + 6 \times 7b$
$$= \quad 6a \quad + \quad 15b \quad + \quad 24a \quad + \quad 42b$$
$$= \quad 6a \quad + \quad 24a \quad + \quad 15b \quad + \quad 42b$$
$$= \quad 30a \quad + \quad 57b$$

When multiplying, we rearrange, and calculate the named numbers arithmetically, the variables according to the methods of indices if they apply.

Example: to simplify $4a \times 3a$
This is equal to $4 \times 3 \times a \times a$
$$= 12a^2$$

Both commutative and associative properties have been used here.

Another example: to simplify $5ab \times 7ab$
This is equal to $5 \times 7 \times a \times a \times b \times b$
$$= 35a^2b^2$$

The next example uses all five properties: which is used where, readers may like to discover for themselves.

$$(a + b)^2 \text{ means the same as } (a + b)(a + b)$$

Initially, regard the second $(a + b)$ as a single number.
The distributive property then gives

$$(a + b)(a + b) = a(a + b) + b(a + b)$$

Now use the distributive property in the usual way.

$$= a^2 + ab + ba + b^2$$
$$= a^2 + ab + ab + b^2$$
$$= a^2 + 1 \times ab + 1 \times ab + b^2$$
$$= a^2 + (1 + 1) \times ab + b^2$$
$$= a^2 + 2ab + b^2$$

This result is often used in reverse, that is,

$$a^2 + 2ab + b^2 = (a + b)^2$$

By giving b a particular value, say, 5, we see that

$$a^2 + 10a + 25 = (a + 5)^2$$

Both of these represent the same number, but for many purposes the right-hand form is more useful than that on the left.

Example: to simplify $(a + 10a + 25)^{\frac{1}{2}}$
This is equal to $[(a + 5)^2]^{\frac{1}{2}}$
$$= a + 5$$

These have been given at full length; with practice, some of the steps may be done mentally. They are offered as illustrations of elementary algebraic calculations. A full account is not possible here, nor is it necessary for the present aim, which is simply to show how the number schema, when combined with the idea of a variable, leads straight into algebra.

Abstracting and embodying

This section could also be called 'mathematical models continued'. Having made use of various mathematical models, it will be useful to make explicit certain activities which have been implicit whenever we have used them, as a preliminary to developing certain algebraic models which are widely used for solving problems.

The first example is a problem of a kind so simple that we solve these almost without noticing. This will allow us to concentrate our attention on the relations between the problem and the model.

Physical problem	*Mathematical model*
There are four of us and three visitors.	
How many places shall I lay at table?	4 + 3 is what?

In going from the left-hand to the right-hand column, we have taken out, or *abstracted*, just those parts of the problem which are relevant to the solution of the problem. This has two advantages. It reduces noise (in the generalized sense; see page 28) and by abstracting only mathematical features it allows us to use a model which we are well practised in working. Going now from right to left,

I must lay 7 places.   ←   7

The answer to the query in the mathematical model is 7. The reversed arrow signifies the process of *re-embodying* this result in the physical realm to give the answer to the original problem. Our method is represented by the diagram below.

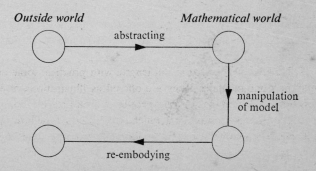

Another problem: a man is quoted £360 as the total cost of a package holiday for himself, his wife and child; knowing that a child is charged half price, how much is the cost per head? This is not difficult either, but it introduces a new aspect of the abstracting process, in that one of the features which we want to abstract is unknown. As a first step towards a mathematical model, we can write

$$\text{adult cost} + \text{adult cost} + \text{child cost} = £360$$

We do, however, know something about the items on the left-hand side. First, each cost is an unspecified number of pounds, so by attaching a unit to a numerical variable we have a mathematical model. Instead of 'adult cost' we can write '£x'. Also, since the child cost is half that of the adult, without knowing what the child cost is we can still represent it by £$\frac{1}{2}x$. Our model now becomes

$$£x + £x + £\tfrac{1}{2}x = £360$$

We are now thinking in Realm 2, that of numbers of units. The abstraction is completed by dropping the units and writing

$$x + x + \tfrac{1}{2}x = 360$$

The use of the letter x for a variable in this context is just a matter of custom. Any letter would do, but it is a common practice among mathematicians to use the letters, a, b, c ... for variables introduced for the second reason given on page 213, that we want to make statements which are true for each and every element in the set, and x, y, z, for variables used for the third reason, that we do not know which element it is, often with the implication that we would like to know.

Models of this kind – equations – are much used in mathematics, usually, of course, for harder problems than this. In the next section we shall develop a schema for further understanding equations and for manipulating them.

Equations and their solutions

What is an equation? This is the name for a particular kind of *statement*. Everyday examples are:

Today is Friday.

The town I live in is Coventry.

The number after 7 is 8.

In each of these statements we are saying that what is written before the word 'is' and what is written after it are different names for the same object. As we saw in Chapter 1, naming classifies, so these statements are not trivial – by telling us that another name may be used, the statement makes available a new classification which may be more useful. 'Coventry' is a more useful name to put at the top of notepaper. 'The town I live in' is more important in the context of, say, rates – what matters to my pocket is the rates of the town I live in, regardless of which town it happens to be.

Since the word 'is' is used with other meanings than the above (compare 'He is running'), we replace it in mathematics by the single-valued symbol ' $=$ ', correctly read 'is equal to', or more conveniently shortened to 'equals'. Here are some mathematical equations:

$$3 + 2 = 5$$
$$7 \times 4 = 28$$
$$101 \text{ (base 2)} = 5 \text{ (base 10)}$$

A mathematical statement, and so an equation, may be either true or false. (In everyday life, statements may be partially true, but in most of mathematics, this is not permissible.) Often, as in the above examples, a statement is made as if it were an assertion, implying the prefix 'It is true that'. Sometimes it is not clear whether this prefix is to be understood or not.

In the case of equations involving variables, such as

$$6x - 3 = 7 + x$$

we cannot say whether they are true or false, since we do not know what number 'x' stands for. To *solve* the equation means to find the set of all values of x for which the equation is a true statement. This *truth set* is called the *solution* of the equation.

Some equations of this kind (equations involving variables) can easily be solved by inspection.

Example: $\qquad\qquad 3 + x = 5$

Solution: $\{2\}$

We can see that this equation is true if x has the value 2 and false otherwise. So the truth set (solution) has only one element. But this is not always the case.

Example:　　　　　$(x - 1)(x - 2)(x + 3) = 0$

If x has the value 1, the first parenthesis $(x - 1)$ becomes $(1 - 1)$ which is equal to zero. Zero multiplied by any other number is zero, so the equation is true in this case: that is, 1 belongs to the truth set. This is also so when x has the values 2 and -3, but does not hold for any other value of x. Therefore

the solution is $\{1, 2, -3\}$

Example:　　　　　$2 + x = 3 + x$

This equation is true for *no* value of x, so the truth set contains no element at all.

Solution: $\{\}$

We call this set 'the empty set' or 'the null set'.

Example:　　　　　$x^2 = 4$

If x represents a natural number, the solution is $\{2\}$.
If however it represents an integer, the solution is $\{+2, -2\}$.

If an equation is not easily solved by inspection, we can sometimes get round the difficulty by solving another equation which has the same truth set. Compare these two statements.

The day after tomorrow = Sunday.
Today = Friday.

If the first of these statements is true, then so is the second. We write:

The day after tomorrow = Sunday　　$\Rightarrow$　　Today = Friday.
(implies)

The reverse implication also applies in this case. If the second statement is true, so is the first.

The day after tomorrow = Sunday　　$\Leftarrow$　　Today = Friday.
(is implied by)

Combining these:

The day after tomorrow = Sunday $\Leftrightarrow$ Today = Friday.

The new sign $\Leftrightarrow$ may be read 'implies and is implied by'. It is, however, an equivalence relation between equations, and the only one in general use, so it is usually read 'is equivalent to'. If either equation is true, then so is the other, and the reader is invited to deduce that if either equation is false, then so also is the other.

By looking at a calendar, it is easy to pick the set of days for which the second statement is true, say, in the month of January 1986. It is 3rd, 10th, 17th, 24th, 31st. This must therefore be the set of days for which the first statement is true, for all those days (and only those days) which make the second statement true will make the first statement true, and vice versa. *Equivalent equations have the same truth set*, so the process of solving the first equation was made easier by finding a simpler equivalent equation.

In everyday life we can usually recognize two equivalent statements, or find another statement equivalent to a given one, intuitively. Mathematical equations, however, become progressively more complex, so we need to find a system. This system, when we have found it, will be yet another schema, of which the 'raw material' will be equations.

Clearly $x = 4$ will be true if and only if x has the value 4. Equally clearly, $x + 1 = 5$ will be true iff x has the value 4.

So $\qquad\qquad\qquad x = 4 \quad \Leftrightarrow \quad x + 1 = 5$

Similarly $\qquad\qquad x = 4 \quad \Leftrightarrow \quad x + 2 = 6$

and whatever number n stands for,

$$x = 4 \quad \Leftrightarrow \quad x + n = 4 + n$$

So an equivalent equation will be obtained if we add the same number to both sides of a given equation. And since the equivalence works both ways, an equivalent equation will also be obtained if we subtract the same number from both sides. By a similar process we can also see that equivalent equations will be obtained if we multiply or divide both sides of a given equation by the same number. The foregoing applies equally whether 'the same number' means a specific number (for example, 5, 7) or a variable (for example, x, $3x$). Simplifying either or both sides by the methods described earlier does not change the number which that side stands for, so clearly this too gives an equivalent equation.

These general principles can be used to obtain progressively simpler equivalent equations from a given equation. These will all have the same solution, for if A, B, C, stand for equations, then

$$A \Leftrightarrow B \text{ and } B \Leftrightarrow C \text{ implies } A \Leftrightarrow C$$

by the transitive property of equivalence relations.

Here is a typical such sequence of equivalent equations, starting with the one given as example earlier.

$\Leftrightarrow 6x - 3$	$= 7 + x$		Add 3 to both sides.
$\Leftrightarrow 6x - 3 + 3$	$= 7 + x + 3$		Simplify.
$\Leftrightarrow 6x$	$= 10 + x$		Subtract x from both sides.
$\Leftrightarrow 5x$	$= 10$		Divide both sides by 5.
$\Leftrightarrow x$	$= 2$		

Solution: $\{2\}$

The last equation of the series is really unnecessary, since we can see from $5x = 10$ that the solution is $\{2\}$.

Note that $x = 2$ is not the solution, though it is often quoted as such. The solution is a truth set, and an equation is not a set but a statement.

We can now apply our equation-solving schema to the equation produced as a model for the 'package holiday' problem in the previous section.

$x + x + \frac{1}{2}x = 360$		Multiply both sides by 2.
$\Leftrightarrow 2x + 2x + x = 720$		Simplify.
$\Leftrightarrow \qquad 5x = 720$		Divide both sides by 5.
$\Leftrightarrow \qquad x = 144$		

Solution: $\{144\}$.

Re-embodying this solution in the original problem, we find that the cost is £144 for each adult and £72 for the child. In general:

Answer(s) to problem. $\leftarrow$ Truth set (solution) of equation.

When solving problems in this way, two tasks are involved: the construction of the mathematical model and its manipulation – often, as here, the solution of an equation. Problems in which both of these tasks are straightforward are usually ones specially constructed as exercises for students, and it is difficult to invent ones which are altogether realistic. Also, the construction of models for events in the physical sciences, such as electronics, space flight, mechanics, requires knowledge of the appropriate science as well as of mathematics, and the initial discovery of the appropriate model has often been the work of a famous scientist. To end this chapter, here are two realistic problems of the latter kind. The model for the first is due to Ohm (1787–1854) and that for the second is based on the discoveries of Newton.

Problem. Wire for the element of an electric fire has an electrical resistance of 25 ohms per metre length. What length of wire should be used for a 1000 watt element, if the mains e.m.f. is 240 volts?

The relation between the power output, the e.m.f., and the total resistance, is embodied in the model.

$$W = \frac{E^2}{R}$$

where W watts is the power output, E volts is the mains e.m.f. and R ohms is the resistance. These, with units attached, are measures of physical qualities; W, E and R by themselves are numerical variables. It is by dropping the units and working in Realm 3, that we can use equations and their associated techniques as models for such a variety of physical situations.

$W = \frac{E^2}{R}$ is a general model. In this problem, W has the value 1000 and E has the value 240. We want to know the corresponding value of R. So the model for this particular problem is:

$$1000 = \frac{240^2}{R} \qquad \text{Multiply both sides by } R$$

$$\Leftrightarrow 1000R = 240^2 \qquad \text{Divide both sides by 1000}$$

$$\Leftrightarrow \quad R = \frac{240^2}{1000} \qquad \text{Simplify}$$

$$\Leftrightarrow \quad R = 57\cdot6$$

Solution of equation: $\{57\cdot6\}$

Re-embodying this in the problem, the total resistance of the wire must be 57·6 ohms. If each metre of wire has resistance 25 ohms, it follows that 2·3 metres of it will be required. (Readers may if they wish make explicit, as an exercise, the simple model used for this last step.)

Problem. At what velocity will an earth satellite be stable in a circular orbit at a height of 1000 km?

The physical qualities which affect the result are the earth's average radius, the distance of the satellite's orbit from the earth's centre, the velocity of the satellite and the gravitational constant. The model which represents the way in which these interact is

$$V^2 = \frac{gR^2}{r}$$

in which R km. is the earth's average radius, r km. the distance of the satellite from the earth's centre, V km. per second the satellite's velocity and g km./sec^2 the acceleration due to gravity. Giving R the approximate value 6400 (the exact value is 6371), r the value 7400 (6400 + 1000), g the value 0·0098, we get:

$$V^2 = \frac{0\cdot0098 \times 6400^2}{7400}$$

$$V^2 = 54$$

Solution: $\{(+7\cdot4), (-7\cdot4)\}$

Re-embodying this solution, the answer to the problem is that the satellite velocity must be 7·4 kilometres per second. The negative value in the mathematical model tells us that an equal velocity in the reverse direction will do just as well.

CHAPTER 13

Mappings and Functions

In Chapter 12, we saw that one way of solving problems in the physical realm is to replace them by a corresponding mathematical problem. The process of abstraction by which we go from the physical to the mathematical realm is however not easy to formulate explicitly. And, given an assortment of problems together with their mathematical models, it may be hard to see anything in common among the abstractive activities by which the models were obtained. These may be reasons why there is (as yet) no generalized mathematics of these abstractive processes.

Within mathematics itself, however, it is often useful to do the same – to replace a problem by another which is easier to solve. And since both the starting and the end points are now exactly stated in mathematical terms, the process of moving from one set of ideas to another within mathematics has the possibility of being studied in at least some cases.

Given two mathematical systems and a correspondence between them which satisfies certain conditions, we have what is called a *mapping*. (The analogy with geographical map-making is not exact, as will be seen shortly.) A well-defined method for getting from one system to another is called a *function*. In this chapter we shall first look at some mappings, and consider what conditions are necessary to qualify for this description. We shall then centre our attention on the functions associated with these mappings,* as entities in themselves, and see whether we can develop operations on functions, in the same way as we can do operations on numbers.

Mappings

The requirements for a correspondence between two systems to be called a mapping are straightforward. Consider first some everyday examples.

---

* Many writers use the terms 'mapping' and 'function' interchangeably, with the meaning here assigned to 'function'.

Suppose we have two sets of objects, {motor cars} and {registration numbers}. In most countries it is required that each car has a registration number, but no car is allowed two registration numbers. Or suppose we have a set of students who have entered for an examination and a set of possible marks. The latter may be {numbers from 0 to 100} or {A, B, C, D, E, F} or {pass, fail}. Then every student must be given one mark, and one only. When laying a dinner table, we lay one place, and one only, for each of the expected diners. Assuming that the above sets are all well-defined, then in each case we have a mapping, the characteristic property being that for each element of the first set (which we will call the original set) there is one, and only one, corresponding element of the second set (which we will call the image set).

What is the difference, if any, between a mapping and the matching process between sets which was described in Chapter 7? For a mapping, every element of the original set must have a corresponding element of the second set. But this does not exclude the possibility that more than one of the Os may share the same X (for example, several students may have the same grade).

Original set   Image set

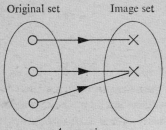

A mapping

On the other hand, there must be no O without a corresponding X (for example, no student without a grade),

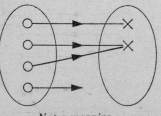

Not a mapping

and there must be no O which has two corresponding Xs (for example, no student must be given two grades).

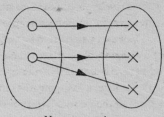

Not a mapping

But it is quite in order for there to be some unused Xs (for example, no student is given grade F).

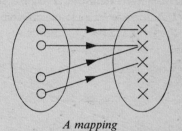

A mapping

The conditions for a mapping are fully satisfied by the matching between sets which we also described as a one-to-one correspondence. (For example, each diner has a place, and only one place. In this case, however, no two diners share a place and there are no spare places.)

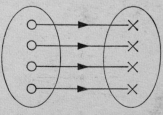

One-to-one correspondence

If we reverse the arrows, we still have a mapping. (For example, each place has a diner, and only one diner. In this case, however, no two places share the same diner and there are no spare diners.) So a one-to-

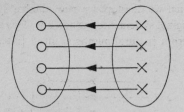

One-to-one correspondence reversed

one correspondence between two sets is a special case of a mapping, with the additional property that it also gives a mapping when reversed. It will be remembered that this reversibility is one of the requirements for any equivalence relation, of which one-to-one correspondence is an example.

It is easy to verify, and the reader is advised to do so, that one-to-one correspondence is the only reversible kind of mapping. In other words, the first two diagrams labelled 'A mapping', which are not one-to-one correspondences, do not give mappings when the direction of the arrows is reversed.

The above diagrams, and the earlier examples, show mappings from one set to another. A mathematical *system* consists of a set, together with one or more operations on the set. (For example, a number system.) So, when mapping from one system to another, we map the elements of the original set to some of the elements of the image set, and operations on the original set to operations on the image set.

We can now see why geographical map-making is not a mathematical mapping, though there is quite a strong resemblance. When making a geographical map, by no means every element in the original set is represented in the image set – indeed, it is doubtful whether the original, which is a certain part of the earth's surface, together with associated qualities such as height above sea level, rainfall, governmental boundaries, is sufficiently well defined to be called a set. An ordinary map is more like a mathematical model, and the method of making it is an abstractive process using physical aids such as levels, theodolites, measures.

Mathematically useful mappings

We have already encountered a series of mathematically useful mappings. At every stage when we developed a new number system there was a mapping of the original system into the new one. This diagram represents the first of these, from {natural numbers} into* {integers}.

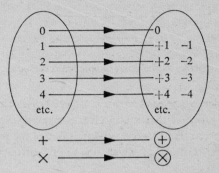

This mapping is not one-to-one, since the negative integers are not images for any of the natural numbers. So it is not reversible. This mapping between {natural numbers} and a sub-set of the integers {positive integers and zero} is, however, reversible. It is, moreover, an isomorphism (see page 187) for both addition and multiplication.

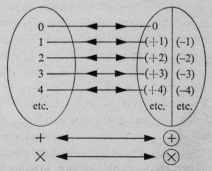

* A note on 'into', 'on to' and 'to'. Mathematicians talk about 'mapping on to' an image set if every element of the image set is the image of some element in the original set, and 'into' if there are some unused elements in the image set. So the term 'into' is correctly used for {natural numbers} → {integers}. I have, however, used 'into' in other places where it would be more accurate to use 'on to', since the distinction does not matter much here. I have also introduced 'to' for individual elements, for example, 'the natural number 2 is mapped to the integer (+2)'. Although I have not seen this done elsewhere, it seems more appropriate than 'into' or 'on to'.

It is the isomorphism at each stage between the earlier system and a sub-set of the new system which makes these particularly useful mappings. For a working model, as distinct from a static one, we need a set with operations on it – a mathematical system; and an isomorphism allows us to work in whichever system we choose, moving freely to and fro between systems.

There is a simple mapping which is often useful when finding averages. Suppose that ten people have heights respectively 161 cm., 180 cm., 172 cm., 175 cm., 190 cm., 163 cm., 176 cm., 160 cm., 169 cm. and 184 cm. To find their average height the straightforward method is to add all these together and divide by ten. But we can reduce the work by subtracting 160 cm. from each height, averaging and then adding 160 cm. to the result.

Original heights in cm.		*Image set*
161		1
180		20
172		12
175		15
190	-160	30
163	$\longrightarrow$	3
176		16
160		0
169		9
184		14

120   Divide this total by 10

$+160$

Result in original set   172 $\longleftarrow$ 12   Result in image set

Implicit in the above is another mapping which is so easy and often used that we hardly notice we are doing it, from a set of measures (in this case, heights in cm.) into a set of pure numbers. Though simple, however, it is non-trivial, and there is more in it than meets the eye, as we saw in Chapter 9.

A mapping which we use when we go abroad is that from one currency into another, for example:

Pounds sterling		Pesetas
10	→	1680
5	→	840
1	→	168
0·50	→	84
0·10	→	16·8

There is also a mapping in the reverse direction, from pesetas into pounds. But since a bank or hotel deducts commission when exchanging currency in either direction, the pairs of numbers which correspond when going from the left-hand column to the right will be different from those which correspond when going from right to left. In this respect, the currency mapping is different from that used for averaging.

A mapping which we use when travelling by train is that between a set of names of stations and a set of times of day.

Manchester	10.00
Stockport	10.09
Wilmslow	10.17
Watford Junction	12.24
Euston	12.44

Here is a mapping which might be of interest to parachutists. It gives the total distance fallen for various times after they jump, before the parachute opens, neglecting air resistance.

Time in seconds		Distance in metres
0	→	0
1	→	4·9
2	→	19·6
3	→	44·1
4	→	78·4
5	→	122·5

Here is another mapping, whose usefulness lies in the rapidity with which manipulations in the image set can be performed, in this case not mathematically but by embodiment in quite a different physical system. The embodiment described below was invented by Hollerith. Though now mainly of historical interest (having been replaced by computers), it still provides a good example for our present purposes, since its physical processes are easier to follow than those of their computerized replacements.

Suppose that an employment agency wishes to choose, from all the people registered with it, those who are graduate mathematicians, speak Italian, are free to travel abroad and are under thirty-five. For each registered person there is a card. The various personal attributes which may be relevant for employment are represented by different locations on the card. Possessing/not possessing one of these attributes is represented by punching/not punching a hole in the corresponding location. So a person who has all the required attributes listed above will be represented by a card with holes punched at all four of the corresponding locations (and possibly at other locations too). In an electrically operated card-sorting machine, hole/no hole can be made to control a switch so that it makes/breaks contact. If these switches are connected in series, it can be arranged for a current to flow and thereby actuate a mechanism which sorts that card out from the rest if, and only if, all the switches are in the make position. These card-sorting machines can work through large numbers of cards at high speed, and can select the sub-set of cards having any required combination of characteristic properties in a very short time compared with that which a human operator would require. In the whole process we can distinguish the following mappings.

{registered persons} → {cards with their names written on}

{relevant attributes} → {locations on cards}

{attributes of same person} → {holes at corresponding locations on
 same card}

{attributes required for a particular job} → {switches connected
 in series with current
 source and electrical/
 mechanical sorting device}

{persons who have all required attributes} → {cards with their
 names on}

{electrical-mechanical selection of cards} → {selection of persons}

The elements of the sets involved include physical objects (persons), written symbols (names), personal qualities (for example, speaking Italian), physical position (of switch contact), operations on certain of the above sets (selecting/rejecting). And the processes by which these mappings are effected are all different – typically, manual with help of typewriter, mental with help of printed symbols, manual or mechanical action of punch, manual via keyboard, electrical–mechanical and, finally perhaps, via a typewritten letter and the post office. We can use whatever method is most convenient, as long as it satisfies the essential requirement that for a given element of the first set there is chosen one and only one element of the second set.

The last example was a mixture of applied mathematics and technology – and, before leaving it, it is worth noticing that not so very long ago, a selection process of this kind could only be done, and on a much more limited scale at that, with the aid of the town crier!

Here, in contrast, is a simple example from pure geometry. It seems intuitively likely that if this figure represents any ellipse* and O is its

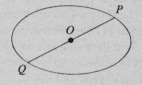

centre, then O is the mid-point of the line segment PQ which passes through O. How can we verify this?

Draw the ellipse in grease pencil on a sheet of glass, and use the sun to project a shadow of this ellipse on a sheet of paper. The paper must be at a right angle to the rays of the sun. By suitably manoeuvring the glass, we can make the shadow of the ellipse into a circle.

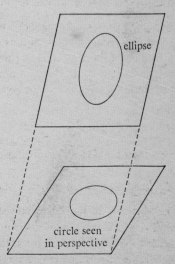

* That is to say, a variable ellipse, an unspecified element of {ellipses}.

From this physical activity a mathematical mapping can be abstracted, and defined in geometric terms, which is called *orthogonal projection*. Here is the original figure and its image by this mapping.

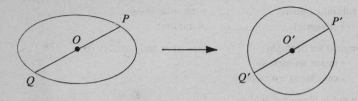

It is easy to prove, by a well-known property of similar triangles, that a line segment and its mid-point project into another line segment with its mid-point. Since we know that in any circle a diameter such as $P'Q'$ is bisected by O' the centre of the circle, we know that in the original figure PQ is bisected by O. And by projection into a circle, many other properties of an ellipse can be found and proved which would be difficult in the original ellipse.

From the foregoing examples, it can be seen that the uses of mappings tend to fall into two classes. In one, we are mainly interested in which element is paired with which (for example, what time the train is at Euston). In the other, we are enabled to side-step a problem by mapping into an image set and solving an easier problem instead.

Functions

For the rest of this chapter, we will turn our attention from functions as parts of mappings to functions as entities in themselves, rather as attention progresses, early in mathematics, from natural numbers as properties of sets to natural numbers as objects of thoughts in their own right. In both cases the process is one of freeing a concept from its origin in any particular set of examples. (See page 57.)

A function is any rule or method whatever whereby, for any and every object in the original set, we can find a (unique) corresponding element in the image set. As we would expect from this description, functions come in many varieties. Looking through some of the mappings described in this chapter, we can find:

Mapping	Function
{motor cars} → {registration numbers}	look in registration book of car
{natural numbers} → {integers}	write + in front of numeral
original set of heights → image set used for calculating averages	omit units and subtract 160
{number of pounds sterling} → {number of pesetas}	multiply by 168
{names of stations} → {times of day}	look at railway timetable
{times in seconds} → {distance fallen by parachutist}	substitute in algebraic formula $d = \frac{1}{2}gt^2$
{points on ellipse} → {points on circle}	a suitable orthogonal projection (by drawing perpendiculars from points in plane of ellipse to image plane)

In the parachutist example, the algebraic formula above represents the function in a different form from the table on page 237. Different ways of symbolizing the same function also help us to understand, or to centre our attention on, different aspects of it. Here is a simple function represented in six different ways.

In words.   the square of

By Venn diagrams
with arrows.

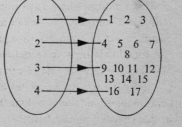

By an equation, in
which x is a variable
element of the original
set and y is the
corresponding element
of the image set.

$$y = x^2$$

By a table.

Original set 1 2 3 4 5 ...
Image set   1 4 9 16 25 ...

By pairing together
corresponding elements
from the two sets.
This gives another set,
whose elements are
ordered pairs.

$$\{(1,1), (2,4), (3,9), (4,16) ...\}$$

By a Cartesian graph.
(The way this is
constructed is described
in Chapter 14, page 254.)

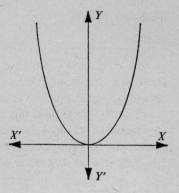

A very useful way is
to show the image of
a variable element
of the original set, thus.

$$x \rightarrow x^2$$

This enables us to specify
various functions by the
different images they give:
here are some examples.

$f:\ \ x \rightarrow x^2$

$g:\ \ x \rightarrow x^3$

$h:\ \ x \rightarrow \dfrac{1}{x}$

These can be read 'f is the function which maps x to x^2', and so on.

In the Venn diagram, the table and the set of ordered pairs, the original set appears to be the set of natural numbers, with zero omitted. But we know that this function can be applied to all kinds of numbers – integers, fractional numbers, rationals, real numbers. Though this might be indicated after a fashion by including a few numbers of this kind in the table (etc.), it is much clearer to state explicitly that *the domain of this function* is {real numbers}. In general, the domain of any function is the set of original objects for which it gives images.

Here are some other examples of functions represented by algebraic equations.

Notice that the domain of the second is {real numbers excluding zero}, since we have no meaning for $\frac{1}{0}$

$$y = 7x + 4$$
$$y = \frac{1}{x}$$
$$y = x$$
$$y = (x + 1)(x - 2)$$

In contrast to the above, which are called *algebraic functions*, here is another kind, all of which have for their domain the set of points in a given plane. Any geometric figure can be thought of* as a set of points, lines being regarded as consisting of points so close together that they are continuous.† So these functions, which are called *geometric transformations*, act on any geometric figure, the image being another geometric figure. In the diagrams below, F represents the original figure with P a variable point on it. F' and P' are their images.

Reflection in the line m, which perpendicularly bisects every line segment such as PP'.

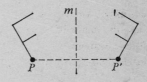

---

* This way of thinking of things is often very convenient mathematically, but it is not intuitively obvious. From everyday experience, lines would be a more direct abstraction, points being regarded as intersections of two lines.

† Just how close together the points have to be we cannot say more clearly without first introducing certain other ideas. Roughly, we mean that there are no gaps.

Rotation about *O*
through a given angle.

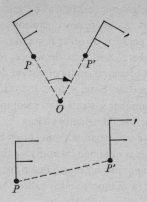

Translation: a movement
without rotation in a
given direction through a
given distance.

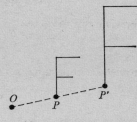

Enlargement with
centre *O*, in this case
by a factor 2.

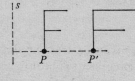

Stretch: an enlargement
in one direction only.
Here, the perpendicular
distance of every point
from the axis of stretch *s*
has been doubled.

Shear: easier to grasp
from the figure than
from words.

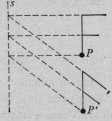

Operations on functions

Here, then, are two major classes of functions: those represented by
algebraic equations like those given as examples earlier, which we will

call algebraic functions; and geometric transformations. It is now natural to ask, are there any operations which we can do on functions like these, and perhaps on other functions, analogous to those which we can do on numbers? And if so, do we get yet another number system? If not, what kind of a mathematical system do we get, and what are its characteristic properties? Since it took four chapters to answer questions of this kind for number systems, and incompletely at that, all that can be given here are a few introductory ideas.

We will begin with algebraic functions. Let us denote the image of x, which is obtained by applying the function f to x, by $f(x)$. Note that this does *not* mean f multiplied by x: it is the result of applying f (a function) to x (a number) and is called the *value of f* for x (or at x). We can say

$$x \text{ is mapped by } f \text{ to } f(x)$$

and abbreviate this to

$$x \xrightarrow{\ f\ } f(x)$$

For convenience in printing or typing, this is often written

$$f: x \to f(x)$$

which may be read as

$$f \text{ maps } x \text{ to } f(x)$$

Since f is an algebraic function which maps numbers to numbers, all its function values will be real numbers. If we now take g to be some other algebraic function, we can define a new function which we call the *sum* of f and g and write as $f \oplus g$, by saying that for all values of x in the domain of both f and g

$$f \oplus g: x \to f(x) + g(x)$$

Here, $\oplus$ refers to addition of the functions f and g, and $+$ refers to addition of $f(x)$ and $g(x)$, which are numbers.

We can also define a function which we call the *dot product* of f and g, written by $f.g$, by saying that

$$f.g: x \to f(x) \times g(x)$$

It is easy to see that, with these definitions, any set of algebraic functions has the properties of a number system. For example,

$$f \oplus g: \qquad x \rightarrow f(x) + g(x)$$
$$g \oplus f: \qquad x \rightarrow g(x) + f(x) = f(x) + g(x)$$
$$\text{since } f(x) \text{ and } g(x) \text{ are numbers.}$$

So addition of functions is commutative, and the other properties for functions under $\oplus$ and . follow similarly from those of real numbers.

These operations are not, however, applicable to the geometric transformations described earlier. Here the domain is a set of points in a plane, the function values are all points too, and we have no meaning for the sum or product of two points. For example, if M stands for the transformation 'reflect in the line m', and if $M(P)$ means the image of a point P under this transformation (which is the same as saying that $M(P)$ is the function value of M for P), and if N and $N(P)$ have similar meanings for reflection in another line n, we have no meaning for $M(P) + N(P)$ analogous to that which we gave to $f(x) + g(x)$.

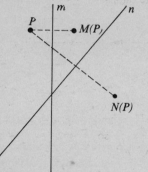

So this particular kind of sum and product are not applicable to functions in general. There is, however, another kind of product which applies to both algebraic functions and geometric transformations, and to many other functions too. This is called the *composition* of two functions, and is the single function which has the same effect as f and g applied in succession. Venn diagrams show it well.

Original set   *Image set for* f   *Image set for* gf

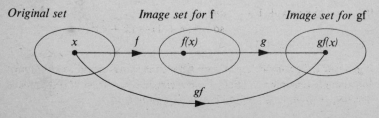

If x is a variable element of the original set, $f(x)$ is the image of x under f and $gf(x)$ is the image of $f(x)$ under g, then gf is the function which maps x all the way to $gf(x)$. In the notation introduced a little earlier,

if	$f: x \rightarrow f(x)$
and	$g: x \rightarrow g(x)$
so that	$g: f(x) \rightarrow g[f(x)]$
then	$gf: x \rightarrow gf(x)$

The way in which the composition product is arrived at explains why it is written gf, although f is the first function to be applied. This product is *not* in general commutative, as an example or two will show.

Example. First, let f and g be algebraic functions, say, f = 'the square of' and g = 'one more than'.

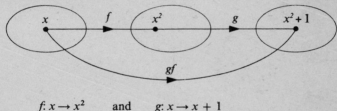

If	$f: x \rightarrow x^2$ and	$g: x \rightarrow x + 1$
	so	$g: x^2 \rightarrow x^2 + 1$
then	$gf: x \rightarrow x^2 + 1$	

Now the other way about.

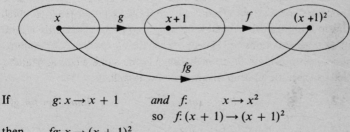

If	$g: x \rightarrow x + 1$ *and*	$f: \quad x \rightarrow x^2$
	so	$f: (x + 1) \rightarrow (x + 1)^2$
then	$fg: x \rightarrow (x + 1)^2$	

A geometrical example. Let M and N have the same meanings as on page 246, namely reflection in lines m and n respectively. Then $NM(P)$ means reflect first in m, result $M(P)$, then in n, result $NM(P)$.

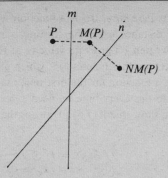

Whereas $MN(P)$ means reflect first in n, result $N(P)$, then in m, result $MN(P)$.

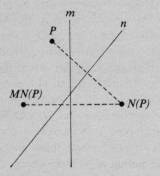

Unless m and n are at a right angle to each other $NM(P)$ and $MN(P)$ are different points.

We have defined three operations on algebraic functions: sum, dot product and composition. Under the first two operations, algebraic functions have the properties of a number system, but these operations cannot be applied to geometric transformations. Under composition, however, many sets of transformations have another mathematical structure, of great generality, called a *group*, which space does not allow us to follow up here.

Differentiation

Those familiar with the calculus will have made frequent use of *differentiation*. This is an operation on a single function which gives another function called its *derivative*. If f is a function which, for some moving object, gives its distance from the starting point at time t, then the derivative of f (written f') gives its velocity at this time; and by differentiating f' we get another function f'', which gives its acceleration.

Differentiating a function for the first time may involve quite a lot of algebra, and to do this every time we need a derivative is inconvenient. But is not hard to show that if $(f \oplus g)'$ means the result of differentiating $f \oplus g$, then

$$(f \oplus g)' = f' \oplus g'$$
$$(f.g)' = f'.g \oplus f.g'$$
and $$(gf)' = g'.f.f'$$

The first of these resembles the distributive property of number systems, and we could say that differentiation is distributive over the sum of functions. The others are new. But these three properties collectively have a similar result to the number system properties, in that by learning the derivatives of a fairly small number of functions, we can obtain quite easily the derivatives of many others, just as by learning a relatively small number of addition and subtraction results, we can add and multiply almost any numbers we wish. We are thereby able to make and manipulate mathematical models whereby many kinds of problems relating to bodies in motion, from high-speed passenger trains to satellites, space flight and the motion of planets, can be solved.

CHAPTER 14

Generalizing some
Geometrical Ideas

Over the preceding chapters, we have taken as our starting point the familiar counting numbers, with the operations of addition and multiplication on them, and have expanded and generalized this basic schema through various number systems, finishing with a look at algebra – to be accurate, the algebra of numerical variables – and functions. In the process we have somewhat neglected geometry, and it is now time to remedy this omission, at least partially. As before, the aim will not be to give the kind of treatment appropriate to a mathematical textbook but to illustrate the process of schematic development and mathematical generalization, this time with geometrical examples. We shall leave on one side the familiar theorems of elementary geometry, about isosceles triangles, tangents to a circle and the like, and shall try rather to find and analyse the most basic ideas of geometry, and then to look at one way in which these can be generalized.

Points in space

'Geometry' means earth measurement, and the Egyptians developed their geometry partly for the purpose of re-establishing land marks and boundaries after the Nile floods. This use of geometry for surveying survives today in the form of trigonometry – earth measurement using a system of triangles as models. Trigonometry is also important for making maps (in the everyday meaning) and for navigation over the sea and through the air. So geometry has its origins as a mathematical model for places and positions, and for movements from one place to another.

Euclidean geometry concentrates on the study of geometric figures, made up of points, lines, triangles, circles, etc., and particularly on the systematic development of the properties of geometric figures from a set of initial assumptions called axioms. It attaches as much importance, or more, to the logical proof of a proposition as it does to the geometric properties themselves; and, because of this, Euclidean geometry was for many centuries regarded as one of the best ways of training the mind in logical thinking. It is interesting to observe that, at present, this attitude among mathematicians is largely reversed, and while geometrical figures are used as aids to the imagination, the final arbiter in questions of logical proof, even of geometrical matters, has become algebra.

As a starting point for mathematical generalization, let us try to find what are the basic elements of geometric figures. These are made up of points, lines (straight and curved), surfaces and solids. However, in Chapter 13 we saw that a line could be thought of as a set of points, and, in the same way, surfaces and solids can also be thought of as sets of points.

Though this way of thinking has certain mathematical advantages, it leaves unanswered the question of what additional qualities a set of points must have before we know it to be, say, a line. One requirement is that it must be an infinite set of points. (Between any two distinct points on a line, we can always insert another. If this were not so, the line would have gaps.) This is not, however, a sufficient condition. Sooner or later in any analysis, we come to a stage where we must say 'look at this, and this, and this', which is to say that we leave the realm of mathematics and enter the realm of sensory experience from which the ideas were first derived. Some pure mathematicians stop short of this boundary, simply saying that these are undefined ideas with relations between them for which no explanation is given. Psychologically this corresponds to detaching the ideas from their origins, which we have already seen to be an important step in generalizing them. To me, this is rather different from ignoring their origins, but this is largely a matter of one's own orientation.

If we agree to look at elementary geometric ideas from the point of view of their origins in experience, we see that when we ignore everything about an object except its location, the corresponding mathematical idea is a point. For example, something as big as a town is often represented by a point on a map. The point represents its position in relation to other geographical features, such as different towns, roads. A road is a way from one town to another – the track of a possible movement. In a map, this is represented by a line, which we can now think of in another way, as the path of a moving point. A movement is a change from one location to another, so whichever way we approach it, points, locations and movements seem to be among the most basic ideas of geometry. It has been suggested that surface and solids can also be thought of as infinite sets of points. The way in which they and lines differ from each other we know has something to do with dimensions, so this is another basic idea about which we must try to become as clear as possible.

We can now list some of the abstractions from which geometry takes its origins, and introduce the idea of a space.

Physical realm	Mathematical realm
Location	Point
Movement	Line
(a connected set of locations)	(a connected set of points)
(the path of a moving object)	(the path of a moving point)
A set of locations	A set of points, which we call *a space*
The shape of an object	A geometric figure – a set of points in a certain configuration

Note that a space may or may not be connected. Most of the ones we are familiar with are connected: for example, those corresponding to the inside of a room or the surface of a table. But we may choose to limit our attention to a set of isolated locations: for example, those at which we can safely emerge from an underground train. The corresponding space is a disconnected set of single points.

Position

We can see by now that, in geometry, position* is all-important. This is perhaps the sharpest point of departure between arithmetic and geometry. The number of a set is independent of the positions of the objects

---

* For convenience we shall refer to the location of a physical object and the position of a point.

in the set, whereas in geometry it is particularly their positions which we are abstracting. Position is, indeed, the only property a point has as such, apart from others which become attached in a particular context (such as being the image of some other point).

When we try to specify the location of objects in the physical world, however, we find that we can only do so relative to that of other objects. These other objects may be near (the walls, floor, door and windows of a room), far (Birmingham is 160 kilometres north-west of London), or imaginary (part of the boundary between the U.S.A. and Canada follows the 49th parallel of latitude). Similarly, in geometry, our only way of specifying points is by their position relative to other points (or lines, which are point sets). A set of points or lines used for this purpose is called a *frame of reference*.

Here is the set of points all of which are 1·5 cm. from *O*. It is an infinite point set, the circumference of a circle.

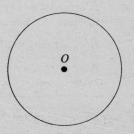

Here is the set of points all of which are less than 1·5 cm. from *O*. It is another infinite point set, the interior of a circle.

Here is the set of points which are 2 cm. from both *A* and *B*. It has two elements *P* and *Q*.

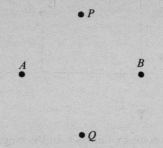

Here is the set of points which are equidistant from both *A* and *B*. It is the line which perpendicularly bisects *AB*. We can only draw part of it.

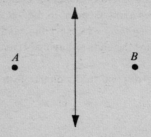

The above are examples of *loci*, point sets which satisfy certain conditions, which can therefore be thought of as the characteristic property of the set. For specifying single points, there are two major systems: Cartesian coordinates and vectors. These will be described in the next two sections.

Cartesian coordinates

This is a method of specifying a point in a plane by two numbers, invented by the French mathematician and philosopher Descartes. The frame of reference consists of two lines, which are usually taken at a right angle to each other. To specify a point *P*, we draw a perpendicular from *P* to *X'X*. Call this *PM* and measure the lengths of the line segments *OM* and *MP*. In the figure above these are 5 units and 3 units respectively. Dropping the units and writing the numbers as an ordered pair (5, 3), we

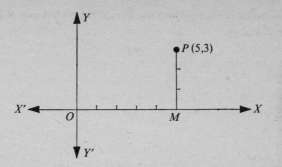

can now specify the point by this pair. In the language of Chapter 13, we can say that Descartes invented a mapping of the set of points in a plane into the set of ordered pairs of real numbers. This mapping is a one-to-one correspondence and is therefore reversible – for any ordered pair of numbers, there is a unique point in the plane. These numbers are called the *Cartesian coordinates* of the point, and since they are pure numbers, we must regard the units as having become part of the frame of reference.

Positive numbers correspond to the directions from O to X and from

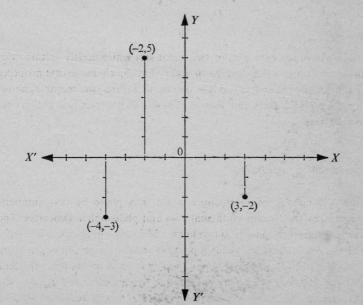

O to *Y*. The diagram at the bottom of page 255 shows some points with negative coordinates. (For convenience, integers have been chosen.) *O* is called the *origin*. Its coordinates are (0, 0).

We have already seen that a function can be represented by a set of ordered pairs. If the ordered pairs are of numbers, the Cartesian mapping enables us to represent a function by a set of points, called a graph. Here are the graphs of the function *the square of* for three different domains.

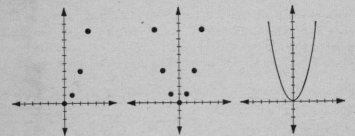

Domain: {natural numbers} Domain: {integers} Domain: {real numbers}

For plotting points, graph paper with a grid of feint lines is generally used. This is difficult to reproduce in print, however. The grid tends to become obtrusive and confuses the figure, so here it has been omitted except on occasions when it is essential. We can also please ourselves whether we insert the letters *X*, *X′*, *Y*, *Y′*, *O*, every time.

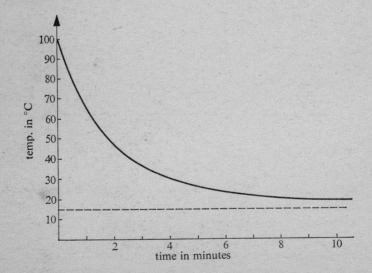

A Cartesian graph gives a good overview of a function and is useful for representing any data which are given as a table of corresponding values. At the bottom of page 256, for example, is a graph called a *cooling curve*.

It shows the temperature of a body (say, a kettle of boiling water) allowed to cool freely in a room at a temperature of 15° C. Moroney's *Facts from Figures*[1] is one of the several texts dealing with descriptive statistics which provide other examples. It can there be seen that the domain of the function need not be a set of numbers for graphical representation. It may be, for example, a set of towns, and the function values may be the mean January temperatures.

But perhaps the greatest breakthrough of the Cartesian mapping results from replacing specific coordinates by numerical variables. In this way, a variable point P of the graph corresponds to a variable pair of coordinates (x, y), where x and y are two real numbers. We show the correspondence briefly by writing $P(x, y)$, meaning the point P whose coordinates are x and y. This links the equation of the function, in this case $y = x^2$, directly with the graph. The graph is the set of all points (x, y) for which the statement $y = x^2$ is true. For example, when (x, y) has the value $(7, 49)$ the equation becomes $49 = 49$, which is true. When (x, y) has the value $(5, 10)$ the equation becomes $10 = 25$, which is false. So $(7, 49)$ lies on the graph, but $(5, 10)$ does not. More briefly we say that $(7, 49)$ *satisfies* the equation, but $(5, 10)$ does not. So the statements (x, y) satisfies the equation and $P(x, y)$ lies on the graph are equivalent.

The usual way to draw the graph of a function is to tabulate any convenient set of pairs which satisfy the equation, plot the corresponding points with the help of squared graph paper, and then (if the graph is continuous) to join the points freehand in a smooth curve. We can also, in the same way, draw the graphs of equations which do not represent functions. Here is the graph of
of $x^2 + y^2 = 25$.
This is not a function,
because for every value of
x which lies between -5
and $+5$, there are two
values of y such that
(x, y) satisfies the equation.
For example, when x has
the value 3, y may have the value $+4$ or -4.

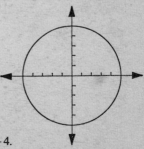

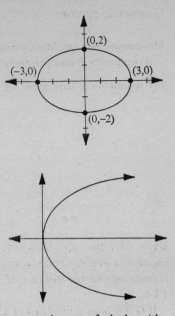

The graph of
$4x^2 + 9y^2 = 36$
is an ellipse through
the points shown.

The graph of
$y^2 = 4x$
is a parabola
(the shape required
for the reflector of a
car headlamp or of a
radio telescope).

This mutual assimilation of the two great schemas of algebra (the algebra of numerical variables) and geometry is among the major achievements of mathematics. It gives us still more opportunities to solve problems in one system by mapping into another; and it helps our thinking by allowing us to symbolize the same ideas in very different ways, the visual-synthetic and the algebraic-analytic. A particular advantage is that it makes available the algebraic methods of the calculus for dealing with geometric problems of gradients, tangents to curves, curvature (whether a graph bends sharply or gently) and areas enclosed by curved boundaries, thereby replacing the approximate methods of drawing and measurement by the exact methods of algebra and numerical calculation.

Space does not, however, allow us to pursue further these two topics of algebraic geometry and calculus. The generalization of geometric space, which is the aim of this chapter, has as its starting point another system of mapping a space which has quite a close resemblance to the Cartesian system, but in which the most important elements are not points but vectors.

Geometric vectors

Movements and locations are closely linked.

If two points P and Q represent two locations, then the line segment *

$\overrightarrow{PQ}$ represents a movement by the shortest path † from P to Q. The

arrow over the top shows that we are talking about a directed line segment: a line segment associated with a direction. The length of PQ corresponds to the distance of the movement (for example, 1 cm. ↔ 1 km.)

If O is a fixed point (part of a frame of reference), we can specify any

point P in a plane by the directed line segment $\overrightarrow{OP}$.

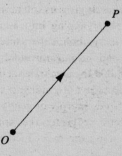

So if our frame of reference also includes a method of specifying directed line segments, we have a system which can represent both locations and movements equally easily.

But what do we mean by a direction?

* A line is regarded as extending indefinitely in both of its directions, so we refer to a 'line segment' when we want to emphasize that we are talking about part of a line.

† We are thinking about locations on a flat surface, not (for example) between London and Sydney or between locations on a curved hillside.

Let this figure represent any rigid object.

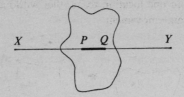

PQ is a line segment fixed in the object and XY is a line fixed in space. If the object moves so that PQ is always part of XY, then we say that the object is *moving in a straight line*.

Now let A,B,C, be three points on the object taken at random.

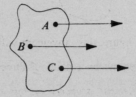

If the arrowed lines through A,B,C, represent their respective directions of motion, do we say that these directions are different? Or do we say that all parts of the object are moving in the same direction? The second view seems more in keeping with common sense, in which case this direction is represented by each or any of the parallel arrowed lines. This means that the geometrical image of a direction is not a single line, but *an equivalence class of parallel arrowed lines*. To say that a line is in a given direction means the same as saying that it belongs to a given equivalence class.

This diagram shows two $\begin{cases} \text{directions} \\ \text{equivalence classes.*} \end{cases}$

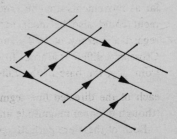

---

* The equivalence relation which determines any of these equivalence classes is ∥ ('is parallel to'), and it can easily be seen that this relation has the three characteristic properties required for an equivalence relation.

To be consistent with the foregoing we must now say that when a rigid object moves from one location to another without turning, all points on it make the same movement.

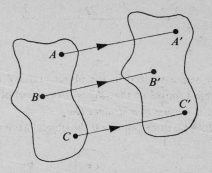

So, in the above figure, this movement will be represented by each and any one of the equivalence class of directed line segments $\overrightarrow{AA'}$, $\overrightarrow{BB'}$, $\overrightarrow{CC'}$. *The characteristic property of this class is called a geometrical vector.*

Since all that follows depends on this idea of a geometrical vector, it is worth pausing a moment here to make sure that the idea has been reasonably well grasped. It is not a hard idea, but may need a little time to sink in. What we are saying, in effect, is that we are regarding all movements of the same distance and direction as being the same movement, ignoring their differences of starting point.

An argument has been put forward to show that this is a reasonable way to look at the matter, but it must in all honesty be admitted that, so far as movements are concerned (though not directions), a good argument could also be made for an alternative viewpoint, in which we regard a journey of 50 km. in the direction NW from, say, Charing Cross as a different journey from one of the same distance and duration from, say, the base of the Eiffel Tower. This would lead us to regard each of the directed line segments $\overrightarrow{AA'}$, $\overrightarrow{BB'}$, $\overrightarrow{CC'}$ as different vectors (though of equal magnitude and direction).

Both of these are perfectly good definitions of vectors, and to distinguish between them we call the first (equivalence class) kind of vector a *free vector* and the second kind a *localized vector*. Free vectors are, however, the most useful mathematically, because they lend themselves more easily to a combining operation (which is given the name of addition). Moreover, we can always turn a free vector into a localized vector by attaching it to a starting point. So hereafter, unless stated otherwise, and generally in mathematical literature, 'vector' means 'free vector' as defined earlier.

Adding geometrical vectors

If a vector represents a movement, a reasonable meaning for adding two vectors which suggests itself is that this should represent the result of combining two movements. Since a vector also specifies a geometric transformation (translation: see page 243), this operation is really an example of the composition of two functions. It is, however, usually called addition, and we shall follow this custom. One possible reason is that this operation is commutative, which composition of functions in general is not. And, as will be seen later, calling it addition fits in rather well with the algebraic way of representing a geometric vector which will lead to a generalized vector. All of which illustrates the point that each new binary operation on a new set is a different entity. Names are given on the basis of a resemblance to some existing operation, but different people may see different resemblances.

Here are two vectors, each an equivalence class of directed line segments representing a movement.

To combine two movements, we begin one where the other one finishes. So to add two vectors, we pick from their equivalence classes any two directed line segments which satisfy this requirement, such as $\overrightarrow{AB}$ and $\overrightarrow{BC}$ on the right of the figure. This makes use of the interchangeability principle for equivalence classes (compare, for example, addition of fractions), which is why free vectors are easier to add than localized ones. Since any member of an equivalence class can be used to represent it, we shall hereafter usually represent a free vector by a single directed line segment, choosing whichever happens to be convenient. It should then be kept at the back of one's mind that this means not just itself, but all in the same equivalence class.

If A represents the start of the first movement, B the end of the first

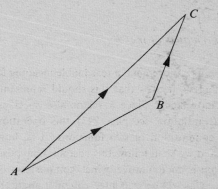

and the start of the second movement, and C the end of the second movement, then the combined effect is that of a single movement which starts at A and ends at C. The directed line segment $\overrightarrow{AC}$ therefore represents the result of adding the two vectors $\overrightarrow{AB}$ and $\overrightarrow{BC}$, and we write

$$\overrightarrow{AB} + \overrightarrow{BC} = \overrightarrow{AC}$$

It is easily seen that addition of vectors as thus defined is commutative. (The diagram for this is at the top of page 264.)

If we complete the parallelogram $ABCD$, since its opposite sides are both parallel and equal in length, these represent the same vector.

So $\overrightarrow{AB} = \overrightarrow{DC}$

meaning that these are two names for the same vector

and $\overrightarrow{BC} = \overrightarrow{AD}$

From the figure, $\overrightarrow{AB} + \overrightarrow{BC} = \overrightarrow{AC}$

and also $\overrightarrow{AD} + \overrightarrow{DC} = \overrightarrow{AC}$

Hence $\overrightarrow{AB} + \overrightarrow{BC} = \overrightarrow{AD} + \overrightarrow{DC}$
$$= \overrightarrow{BC} + \overrightarrow{AB}$$

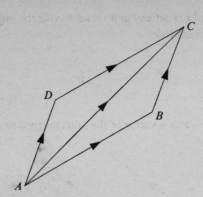

Multiple of a vector by a number

Let us now use small letters in bold type, thus: **a**, **b**, **c**, to denote geometric vectors.

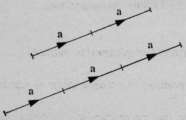

Then **a** + **a** is another vector in the same direction as **a**, but represented by a line segment twice as long. We say that **a** + **a** has twice the *magnitude of* **a**, and write:

$$\mathbf{a} + \mathbf{a} = 2\mathbf{a}$$

Similarly

$$\mathbf{a} + \mathbf{a} + \mathbf{a} = 3\mathbf{a}$$

and so on. In this multiplication, the two things being multiplied are of different kinds, the first being a number and the second being a vector. This 'having a foot in two camps' is one of the major features of geometric vectors which is kept when generalizing.

From multiplication of a vector by a natural number, we can easily generalize to multiplication by any real number. First, let k be a positive real number. Then $k\mathbf{a}$ is defined as a vector having the same direction as $\mathbf{a}$ and k times its magnitude.

Next, we define $-\mathbf{a}$ as a vector of the same magnitude as $\mathbf{a}$ but in the oppposite direction.

Adding in the usual way gives $\mathbf{a} + (-\mathbf{a}) = 0$

which justifies the definition.

If k is a negative real number, then $-k$ is a positive real number, so we can define $k\mathbf{a}$ as equal to $(-k)(-\mathbf{a})$: that is, a vector in the opposite direction to $\mathbf{a}$ and $(-k)$ times its magnitude.

Frame of reference for geometric vectors

We are now in a position to set up a frame of reference for geometric vectors.

First, we choose two perpendicular directions. Any vector can be resolved (split up) into two vectors, one in each of these directions. All we have to do is draw a line in one direction through one end, and a line in the other direction through the other end.

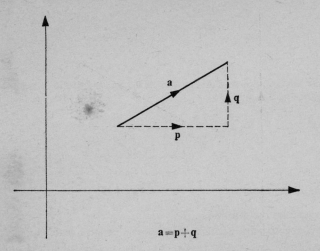

$$a=p+q$$

Next we choose a unit of length and let **i** and **j** be vectors of unit magnitude in these directions. **p** and **q** can now be expressed in terms of these unit vectors, since we can find real numbers x and y such that **p** = x**i** and **q** = y**j**.

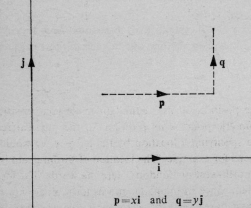

$$\mathbf{p}=x\mathbf{i} \quad \text{and} \quad \mathbf{q}=y\mathbf{j}$$

Combining these steps, we can write any vector a in the form x**i** + y**j**, where x and y are real numbers and **i** and **j** are unit vectors in perpendicular directions, called the *base vectors*.

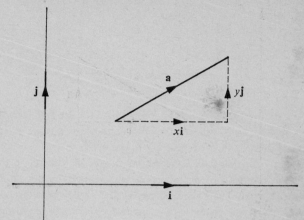

This completes our frame of reference for geometric vectors. We can very easily extend it into a frame of reference for points in this space by adding a point of origin O. (The position of) the point P is then specified by the vector $\overrightarrow{OP}$, which is called the *position vector* of P relative to O.

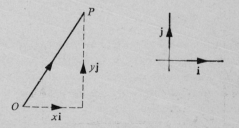

The words 'the position of' are redundant in the mathematical realm because a geometric point *is* its position. In the physical realm, this corresponds to specifying a location by the journey we would have to make to get there: for example, 'The location of Birmingham is 160 km. in a direction north-west of London.' Here the words 'The location of' are not redundant, since Birmingham and its location are not the same. A combined frame of reference for vectors and points is often drawn like that on page 268.

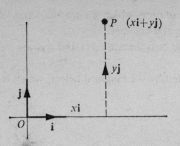

This is all right as long as we do not think that the unit vectors are localized at O, and it shows well the close connection between the vector frame of reference and Cartesian coordinates: (x, y) and $(x\mathbf{i} + y\mathbf{j})$ are different ways of naming the same point, and, of the two, the Cartesian system is the simpler. For dealing with sets of points in the plane, it is generally preferred. The difference lies in the ability of the second to represent vectors, and manipulation thereof, as we shall see in the next section.

Addition of vectors, and multiplication by a number, can now be done algebraically.

Let $\quad \mathbf{a} = x_1\mathbf{i} + y_1\mathbf{j} \quad$ and $\quad \mathbf{b} = x_2\mathbf{i} + y_2\mathbf{j}$
Then $\mathbf{a} + \mathbf{b} = x_1\mathbf{i} + y_1\mathbf{j} + x_2\mathbf{i} + y_2\mathbf{j}$
$$= x_1\mathbf{i} + x_2\mathbf{i} + y_1\mathbf{j} + y_2\mathbf{j} \qquad \text{(step 1)}$$
$$= (x_1 + x_2)\mathbf{i} + (y_1 + y_2)\mathbf{j} \qquad \text{(step 2)}$$

In step 1, the rearrangement of the terms depends on the commutativity of addition of vectors, which we have already demonstrated. Step 2 makes a further assumption, of a mixed sort of distributive property. There is more here than meets the eye, because the $+$ in $(x_1 + x_2)$ means addition of the numbers x_1 and x_2, whereas the $+$ in $x_1\mathbf{i} + x\mathbf{i}_2$ means addition of the vectors $x_1\mathbf{i}$ and $x_2\mathbf{i}$, which is a different operation.

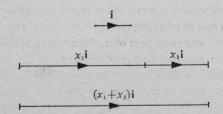

The conventional notation, which we are following, obscures this difference but makes the algebra easy.

The result is true only because $x_1\mathbf{i}$ and $x_2\mathbf{i}$ are vectors in the same direction.

Step 1 is illustrated by the diagram below, which is also suggestive of step 2.

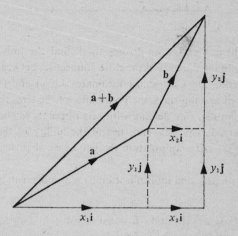

To find the algebraic method for multiplying a vector by a number, we need first a theorem about triangles.

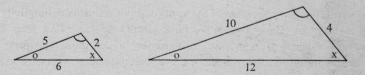

In these two triangles, corresponding angles (marked alike) are equal in size. Also, the ratios 2:4, 5:10, 6:12, are in the same proportion. The theorem and its converse together tell us that if one of these properties is true for a given pair of triangles, then so is the other. These triangles are said to be *similar*, and since each of the properties implies the other, we can choose either as our definition of similarity.

To apply this to vectors, and remembering that $k\mathbf{a}$ means a vector k times as long as $\mathbf{a}$ in the same direction, we draw the two triangles overleaf.

Both of these have their corresponding sides in the same three directions, and so their corresponding angles are equal in size. Hence, the triangles are similar, and we can fill in the lengths of the other sides of the second triangle from our knowledge of equivalent ratios.

$k\mathbf{a} = kx\mathbf{i} + ky\mathbf{j}$

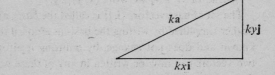

This is what we would have hoped for, from an algebraic point of view. But there is a hidden assumption here too. All we know so far is that $k\mathbf{a} = k(x\mathbf{i}) + k(y\mathbf{j})$, in which the parentheses mean that we are multiplying the vector $x\mathbf{i}$ by the number k, and similarly for the other. If these are specified numbers, say, $3 \times (2\mathbf{i})$, it is tempting to simplify this as $6\mathbf{i}$, assuming a mixed associativity, that $3 \times (2\mathbf{i}) = (3 \times 2)\mathbf{i}$. Indeed, it may seem over-cautious, even pedantic, to question the point, but when working with a system consisting partly of numbers and partly of quite different things, called vectors, we need to be careful. Part of the mathematical way of thinking consists in looking for, and checking, these hidden assumptions, in the same way as, if we are wise, we read carefully the small print as well as the glowing descriptions before we sign on the dotted line.

This represents $3 \times (2\mathbf{i})$

This represents $(3 \times 2)\mathbf{i}$
$= 6\mathbf{i}$

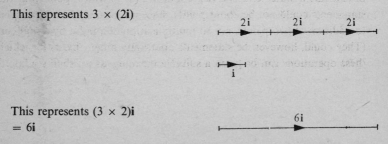

and it is clear that this method is general. So if $k(x\mathbf{i}) = (kx)\mathbf{i}$, we can write $kx\mathbf{i}$ to mean either or both.

Basis

Algebraic addition of vectors and multiplication by a scalar can be done more simply if we separate the real numbers x and y, which specify a particular vector, from the base vectors $\mathbf{i}$ and $\mathbf{j}$, which specify the frame of reference.

Instead of $x\mathbf{i} + y\mathbf{j}$ we write $(x, y)\{\mathbf{i}, \mathbf{j}\}$.

The set of base vectors $\{\mathbf{i}, \mathbf{j}\}$ is called the *basis*, and the notation can be further simplified by writing the basis as $\mathscr{B}$; or, if the frame of reference is known and does not change, by omitting it altogether. So addition of two vectors can now be written in any of these ways, all of which mean the same.

$$(x_1\mathbf{i} + y_1\mathbf{j}) + (x_2\mathbf{i} + y_2\mathbf{j}) = (x_1 + x_2)\mathbf{i} + (y_1 + y_2)\mathbf{j}$$
$$(x_1, y_1)\{\mathbf{i}, \mathbf{j}\} + (x_2, y_2)\{\mathbf{i}, \mathbf{j}\} = (x_1 + x_2, y_1 + y_2)\{\mathbf{i}, \mathbf{j}\}$$
$$(x_1, y_1)\mathscr{B} + (x_2, y_2)\mathscr{B} = (x_1 + x_2, y_1 + y_2)\mathscr{B}$$
$$(x_1, y_1) + (x_2, y_2) = (x_1 + x_2, y_1 + y_2)$$

Multiplication by a number:

$$k(x\mathbf{i} + y\mathbf{j}) = kx\mathbf{i} + ky\mathbf{j}$$
$$k(x, y)\{\mathbf{i}, \mathbf{j}\} = (kx, ky)\{\mathbf{i}, \mathbf{j}\}$$
$$k(x, y)\mathscr{B} = (kx, ky)\mathscr{B}$$
$$k(x, y) = (kx, ky)$$

By omitting reference to the basis altogether, as in the last line of each set of equivalent statements, we risk confusing (x, y) with a point. But the statements could not be about points, since no meaning has been given to addition of two points, nor to multiplication of a point by a number. (They could, however, be statements about any other objects for which these operations can be given a suitable meaning, as we shall see later.)

Dimensions

So far we have been considering geometrical vectors in a plane, which is a space of two dimensions. The number of base vectors is also two. Is this coincidental, or are these numbers necessarily equal?

Suppose we take three base vectors **p**, **q**, **r**, with **p** and **r** at a right angle and **q** at 30° to **p** and 60° to **r**. Then any given vector **a** can always be resolved into multiples* of these vectors.

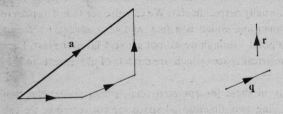

But this resolution is not unique.

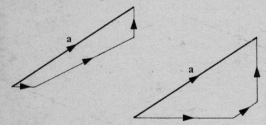

And each of the base vectors can be expressed as the sum of multiples of the other two. In each case, one of the multiples is negative, as shown by the reversed direction.

Clearly we cannot express all vectors in a plane whatever their direction as the sum of multiples of a single vector. We have also seen that if we

---

* Hereafter we shall not bother to say 'multiple by a number,' since multiple of a vector by a vector has not here been given a meaning.

take three vectors we can replace one of them by the sum of multiples of the other two, and thereby reduce the number of base vectors to two. So it seems that in this case the number of dimensions of the space is the *least* number of *independent* base vectors which will form a basis. And by a basis we mean a set of base vectors such that any vector in the space we are considering can be expressed *uniquely* as the sum of multiples of these vectors.

It is easy to extend the above arguments to a geometrical space of three dimensions, for which a convenient basis is a set of three unit vectors mutually perpendicular. We can also see that it applies to a space of one dimension, which is a line, and to a space of zero dimension, which is a point – though we cannot get very far in the last. These are the only geometrical spaces which are models of movements in the physical realm.

For any of these (except zero dimensional space) the basis is not unique. Using two-dimensional space for convenience, we can choose pairs of unit vectors in different directions.

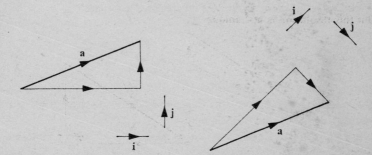

There is no difficulty in resolving **a** into vectors having the directions of **i** and **j** in either of these cases. We can change the sizes of the unit vectors – this will simply change the number we have to multiply them by, in a particular case. For the same reason they need not be of equal magnitude: in geometrical vectors it is often convenient to make them so, but in generalized vectors we shall see that the units may be of quite different kinds. And, so far as geometrical vectors are concerned, they need not even be perpendicular. This was something we took over from the physical realm, in which movements have to be perpendicular to be independent. (We can go north without making any movement east, but if we go north-east we are also going east.) To generalize the idea of independence so that it applies also to non-perpendicular vectors is not difficult, but we do not particularly need it for the next step, so we shall not digress here to do so.

All the foregoing statements are about free vectors, and the frame of reference (basis) also consists of a set of free vectors. Points are no longer the most important objects, though we can include them if we wish by including a point of origin in the frame of reference, other points then being specified by their position vectors relative to the origin.

Let us summarize the chief properties of our new space. It consists of mathematical ideas called vectors, which took their origin from physical movements but have now become free of this origin. These can be added and multiplied by numbers, according to certain rules. Any vector in a given space can be uniquely expressed as the sum of multiples of members of a set of base vectors. This set of base vectors is called a basis for the given space. A basis for a space is not unique, but its vectors must be mutually independent, meaning that we cannot express any of them as the sum of multiples of the others. The number of vectors in any basis of a space is equal to the number of dimensions of a space.

Generalized vectors

Having formulated the properties of geometrical vectors independently of their origins, we can find other things which also have these properties. Compare the following two examples.

A set of geometrical vectors has as basis $\{i, j, k\}$, where these are mutually perpendicular unit vectors.

If $\qquad\qquad a = (2, 3, 5)\{i, j, k\}$
and $\qquad\qquad b = (4, 6, 2)\{i, j, k\}$
then $\qquad\qquad a + b = (6, 9, 7)\{i, j, k\}$
and $\qquad\qquad 5a = (10, 15, 25)\{i, j, k\}$

A set of households can be represented by the same notation with basis $\{m, w, c,\}$, where m = man, w = woman, c = child.

As before, if $\qquad h = (2, 3, 5)\{m, w, c\}$
$\qquad\qquad\qquad g = (4, 6, 2)\{m, w, c\}$
then $\qquad\qquad h + g = (6, 9, 7)\{m, w\, c\}$
and $\qquad\qquad 5h = (10, 15, 25)\{m, w, c\}$

How far does the second system behave like the first? 'Adding' house-holds makes sense – they just move in together. So does 'multiplying by a number', in this case by 5. This means that 5 equal households move in together. At first sight, the basis does not quite fit the pattern. The basis of a set of geometric vectors is itself a set of vectors, so the basis of this space of households ought to be a set of households. This is easily taken care of. We define m as a household consisting of one man, and likewise for w and c. So now

$$m \text{ is short for } (1, 0, 0)\{m, w, c\}$$
$$w \text{ is short for } (0, 1, 0)\{m, w, c\}$$
$$c \text{ is short for } (0, 0, 1)\{m, w, c\}$$

To write the first household h as the sum of multiples of m, w and c, we see that

$$h = 2 \times (1, 0, 0)\{m, w, c\} + 3 \times (0, 1, 0)\{m, w, c\} +$$
$$5 \times (0, 0, 1)\{m, w, c\}$$
$$= (2, 0, 0)\{m, w, c\} + (0, 3, 0)\{m, w, c\} + (0, 0, 5)\{m, w, c\}$$
$$= (2, 3, 5)\{m, w, c\}$$

This justifies the way we first wrote the basis, as simply $\{m, w, c\}$, and allows us to go on using this as a convenient shorthand.

Our requirement that m, w and c shall be independent is also satisfied. Clearly if we add to any household another consisting only of a single man, it makes no change to the number of women and children. Mathematically, we cannot express m as the sum of multiples of w and c, and the same is true for the other two. If somebody notices that the category 'child' is not well-defined (which also affects the other two), we can make it so: say, a human of either sex who has not yet reached his or her sixteenth birthday.

Notice also that there is no implied operation of 'adding' men, women and children. $(2, 3, 5)$ $\{m, w, c\}$ means 2 men *and* 3 women *and* 5 children where 'and' here means 'living in the same house as'. This is consistent with what was pointed out earlier on page 268, that adding has different meanings for vectors and for numbers.

Dropping the word 'geometrical' in front, we have made out a good case for regarding the set of households as another example of a set of vectors, generalizing this concept in the same way as we did the idea of number and of powers of a number. This we shall now do, without stopping to dot every i and cross every t, since there are altogether rather a lot of them – the important ones have been or will be mentioned, and a complete list will be given at the end of the chapter, for reference rather than essential reading.

We can think of 'household' vectors with more than three dimensions, since we are no longer bound by the limitations of geometrical space, generalizing now the idea of dimension also. For example, we can separate the children into boys and girls, which gives a set of four-dimensional vectors. We may also be interested in how many automobiles, television sets and cameras belong to each household. This gives us seven-dimensional vectors in which $h = (2, 3, 5, 0, 3, 1, 4)\{m, w, b, g, a, t, c\}$ represents a household consisting of 2 men, 3 women, 5 boys, 0 girls, 3 automobiles, 1 television set and 4 cameras. It is easily seen that these new vectors have all the properties of the former ones.

In the above example, the numerical multipliers have to be natural numbers. We cannot assign a meaning to negative, or fractional, numbers of men, women, children, cars, etc. Can we find some vectors to which this restriction does not apply?

Suppose that a man likes to keep his money in a number of different banks. We need not say how many – call this number n. Then, if 'x_1b_1' means that he has x pounds in the first bank, etc., his overall banking state can be represented by the n-dimensional vector $X = (x_1, x_2, x_3 \ldots x_n)\mathscr{B}$, where $\mathscr{B}$ is short for the set of his banks $= \{b_1, b_2, b_3, \ldots b_n\}$, and also, of course, for basis.

Assuming that his bank accounts can be overdrawn, $x_1, x_2, x_3 \ldots$ can now be positive, zero or negative integers. (If he is not to be infinitely rich or in debt, some restriction will still be necessary.) Addition of two banking vectors X and Y, defined as the amalgamation of corresponding bank accounts of two persons, follows the established method. Multiplication of a banking vector by a positive integer is straightforward, and so is multiplication by zero, though possibly disastrous. Multiplication by a negative integer depends on our being able to find a meaning for $-X$, which is harder. Suppose, however, we look at things from the point of view of a bank. Then, if the man has £1000 in credit, the bank owes him £1000, and vice versa. So $-X$ could be given the meaning of 'look at his account from the opposite side of the counter'.

For an example in which the multipliers can be not only integers but rational numbers, let $v_1, v_2 \ldots v_n$ be the electric potentials at points P_1,

P_2 ... in an electrical network, due to a given current input I. We may represent this overall state by an n-dimensional voltage vector $V = (v_1, v_2 \ldots v_n)\mathscr{B}$. The basis here is the set {volts at P_1, volts at P_2 ...} If V' is the voltage due to a current I', then $V + V' = (v_1 + v_1', v_2' + v_2 \ldots v_n + v_n')\mathscr{B}$; and here I' can be in the opposite direction to I, in which case it will have the opposite sign. The voltage vector due to a current kI is $V = (kv_1, kv_2 \ldots kv_n)\mathscr{B}$. Here k can be a positive or negative rational number, a negative multiplier representing reversal of the direction of the current. If we are not concerned with actually measuring the current and voltages, there is no reason why we cannot allow k to be any real number.

So far, each generalization of the idea of a vector has been accompanied by an example from the physical realm. Examples of this kind should be regarded as aids to conceptualization rather than mathematical justifications – though if we set up a system of which we cannot find any examples, it will be more difficult to think about. The last examples have prepared the way for a further generalization, in which we drop the basis altogether and simply regard a set of vectors as a set of ordered n-tuples of the form:

$$X = (x_1, \qquad x_2 \qquad \ldots \qquad x_n)$$
$$Y = (y_1, \qquad y_2 \qquad \ldots \qquad y_n), \text{ etc.}$$

such that $X + Y = (x_1 + y_1, x_2 + y_2 \ldots x_n + y_n)$
and $\qquad kX = (kx_1, \qquad kx_2 \qquad\qquad kx_n)$

We have, in dropping the basis, now given up any geometric or physical interpretation, and turned our vectors into purely mathematical objects. Further than this stage we shall not go, partly because examples needed would become increasingly mathematical, requiring quite a lot of preliminary explanation, but largely because the aim of this chapter should by now have been achieved: that of illustrating the process of mathematical generalization in a geometrical context.

In the process, however, the ideas seem to have become almost entirely algebraic. This is true not only for vectors, but for geometry in general. One opens a book on advanced geometry and finds it full of statements in algebraic form, possibly quite empty of geometric figures. Is this inevitable? I do not know. It may be, but the question needs to be asked. Spatial thinking is a mode of functioning of intelligence which, it was suggested in Chapter 5, may be complementary to verbal. The great success with which the Cartesian system represents algebraic ideas (and particularly those of calculus), shows them in a different light, and gives a synoptic view, suggests that there could still be a renaissance of an area of mathematics for which we now lack a name, but which would depend more on the visual apprehension of relationships than on verbal-algebraic reasoning.

Vector spaces*

In the last section, though the household and banking vectors were fairly described as such, the systems as a whole fell short of the total requirements of a *vector space*, and care was therefore taken to refer to them simply as sets of vectors on which certain operations were defined. The purpose of this section is simply to list and explain, for those who are interested, what these requirements are.

Within the set of vectors themselves there are five necessary properties. If X, Y, Z, are any vectors in the set, it is required that

1. $X + Y$ is defined and also belongs to the set.
2. $X + Y = Y + X$
3. $(X + Y) + Z = X + (Y + Z)$
4. There is an element 0 called the zero element such that
$$X + 0 = X \text{ for all } X \text{ in the set.}$$
5. For every X in the set, there is another element $-X$, also in the set, such that
$$X + (-X) = 0$$

Property 1 is that of closure, required for any binary operation on a set. The next two say that addition must be commutative and associative. The set of household vectors satisfies all of these, and also 4 (we can have an empty house), but not 5. The set of banking vectors satisfies all of these requirements if we set no limit to how much he may be in credit, or overdrawn; otherwise we might get into difficulty with 1.

The chief shortcoming of these two systems is, however, in the restriction of the multipliers to natural numbers and integers respectively. For a vector space these must be rational numbers, real numbers or any

---

* This is mainly a tidying-up section, and may be omitted without detriment to the central theme.

other system of a kind which is called a *field*. What a field is we shall not here describe in detail. Briefly, the rational and real number systems have certain extra properties which the natural, fractional and integral number systems do not have, and which therefore go beyond the minimum of those required for a number system. It is this longer list of properties which characterizes a field. So the rationals and reals are fields; but there are other systems which we have not mentioned, which are also fields. Conditions 6–10 refer to the relation between the set of vectors and the elements of the field. We still call the mixed operation on one element of the field and one element of the set of vectors multiplication, and require that if X, Y, are any vectors of the set and α, β, are any elements of the field, then

6. αX is defined and belongs to the set of vectors.
7. $(\alpha + \beta) X = \alpha X + \beta X$
8. $\alpha(X + Y) = \alpha X + \alpha Y$
9. $\alpha(\beta X) = (\alpha\beta)X$
10. $1X = X$ where 1 is the unity element of the field.

All of these requirements would be satisfied by both household and banking vectors if α and β could be *any* rational numbers. We can therefore use either of these examples to illustrate their meanings. 6 says that multiplying a household by a number gives a household (not a number, or anything else). 7 and 8 describe two kinds of distributivity. The first says that 'mixed multiplication' is distributive over addition of numbers, while the second says that it is distributive over addition of vectors. Both of these can easily be verified in the examples of household and banking vectors. Condition 9 was discussed earlier, on page 270. The last one looks trivial, but is not. In any number system, we have an element which, if we multiply a number by it, gives the same number as before. In the systems we know, this number is respectively 1, $(+1)$, $\frac{1}{1}$, $(+1):(+1)$ and $1\cdot000\ldots$ One of the requirements for a field is that it too has an element of this kind, which is called the unity element, since it is a generalization of the idea of unity. What Condition 10 says is that the unity element for multiplication within the number system or other field which defines a vector space also acts as a unity element for the mixed multiplication.

Retrospect

We have come a long way since the beginning of Chapter 8. Most of the going has been uphill, using the metaphor here to mean that abstractions have repeatedly been built on abstractions, nearly always moving towards higher abstractions. Moreover, little time has been spent at each level before setting off for the next, which, as explained earlier (page 132), is not the best way to travel if time allows a more leisurely approach. In a progress of this kind, one tends to concentrate, from necessity, on ground nearby. Now it is time to look back over the ground which has been covered and try to get an overview of the whole.

This can best be done by schematic diagrams in which conceptual hierarchies are represented by spatial arrangements on paper. Having drawn a diagram of this kind for the whole of Chapters 7 to 14 taken together, I decided that it was too complicated to be useful and replaced it by the series of diagrams which follow – one for each chapter, except that Chapters 9 and 10 are taken together. The diagrams read from the bottom upwards. Parentheses mean that an idea has already been developed in a previous chapter. If a study of each diagram is combined with a quick reference back in the text to any ideas which have temporarily been lost sight of – metaphorically, taking a closer look through binoculars, while keeping one's present standpoint – the reader may hope to be able, to some degree at least, to see both the forest and the trees. Having done this, it might be a useful conclusion to take a large sheet of paper and combine all these diagrams into one, for a diagram which is too complicated to be useful when first encountered in a finished state becomes much more intelligible if one has drawn it oneself.

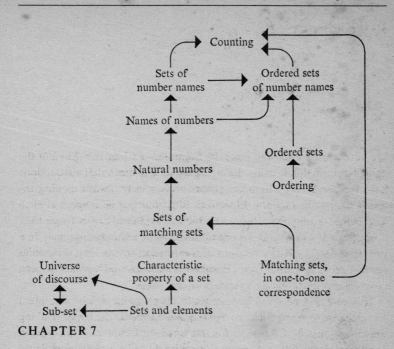

CHAPTER 7

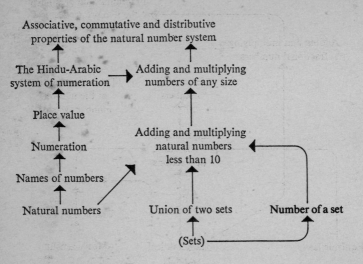

CHAPTER 8

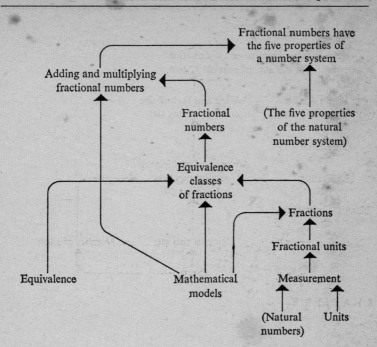

CHAPTERS 9, 10

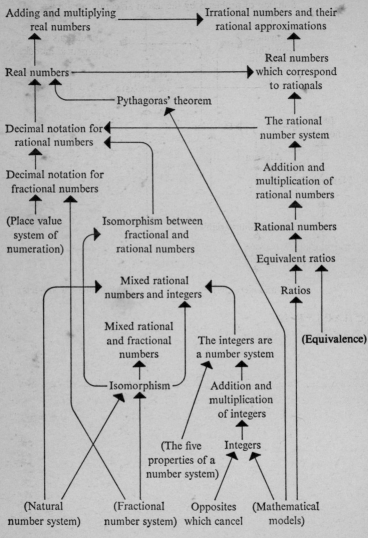

Adding and multiplying real numbers

Irrational numbers and their rational approximations

Real numbers

Real numbers which correspond to rationals

Pythagoras' theorem

Decimal notation for rational numbers

The rational number system

Decimal notation for fractional numbers

Addition and multiplication of rational numbers

(Place value system of numeration)

Isomorphism between fractional and rational numbers

Rational numbers

Mixed rational numbers and integers

Equivalent ratios

Ratios

Mixed rational and fractional numbers

The integers are a number system

(Equivalence)

Isomorphism

Addition and multiplication of integers

(The five properties of a number system)

Integers

(Natural number system)

(Fractional number system)

Opposites which cancel

(Mathematical models)

CHAPTER 11

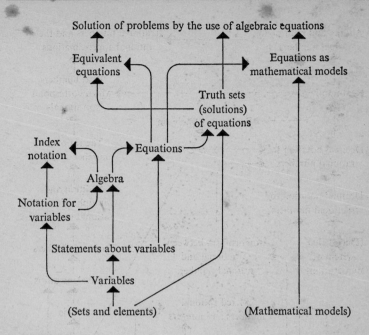

CHAPTER 12

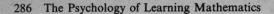

Geometric
transformations

Sum and
dot-product of
two functions

Composition of
two functions

Mappings and functions

Two sets

CHAPTER 13

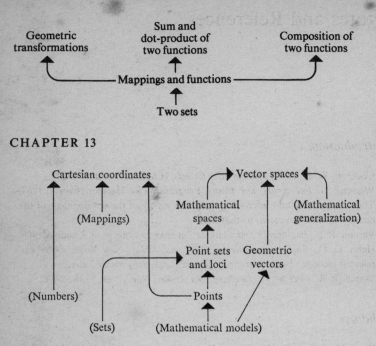

Cartesian coordinates

Vector spaces

(Mappings)

Mathematical
spaces

(Mathematical
generalization)

(Numbers)

Point sets
and loci

Geometric
vectors

Points

(Sets)

(Mathematical models)

CHAPTER 14

Notes and References

Introduction

1. Cockroft, W. H., *et al.*, *Mathematics Counts*, H.M.S.O., London, 1982.
2. Wiseman, S., *Intelligence and Ability*, Penguin Books, Harmondsworth, 1967. This is a valuable and interesting historical survey of the development of the concept of intelligence, up to that date.
3. Vernon, P. E., *Intelligence and Cultural Environment*, Methuen, London, 1969.
4. Hebb, D. O., *The Organisation of Behaviour*, Wiley, New York, 1949. The relevant passage also appears in Wiseman's anthology already cited.
5. Skemp, R. R., *Intelligence, Learning, and Action*, Wiley, New York, 1979.

Chapter 1

1. Vygotsky, L. S. (trans. Hanfmann, E., and Vakar, G.), *Thought and Language*, M.I.T. Press, Cambridge, Massachusetts, 1962.

Chapter 2

1. I mean here relational understanding. See Skemp, R. R., 'Relational understanding and instrumental understanding', *Mathematics Teaching* 77, 1976.
2. This experiment is reported in full in Skemp, R. R., 'The need for a schematic learning theory', *Brit. J. Ed., Psychol.*, xxxii, 133–42, 1962.
3. Bell, E. T., *Men of Mathematics*, Chapter 19, Penguin Books, Harmondsworth, 1937.
4. Bell, M. A., Unpublished M.Sc. Thesis, University of Manchester, 1967.

Chapter 3

1. Piaget, J., *Judgement and Reasoning in the Child*, Routledge & Kegan Paul, London, 1928.

2. Inhelder, B., and Piaget, J., *The Growth of Logical Thinking*, Routledge & Kegan Paul, London, 1958.

Chapter 4

1. Ghiselin, B. (ed.), *The Creative Process*, University of California Press, Berkeley, 1952, and also Mentor, New York, 1955.

Chapter 5

1. Hadamard, J., *The Psychology of Invention in the Mathematical Field*, Dover, New York, 1945.
2. Macfarlane Smith, I., *Spatial Ability*, University of London Press, London, 1964.
3. Glennon, V. J., *Neuropsychology and the Instructional Psychology of Mathematics*, Research Council for Diagnostic and Prescriptive Mathematics, Kent, Ohio, 1980.

Chapter 7

1. Piaget, J., *The Child's Conception of Number*, Routledge & Kegan Paul, London, 1952.

Chapter 14

1. Moroney, M. J., *Facts from Figures*, Penguin Books, Harmondsworth, 1951.

Suggestions for further reading

BARTLETT, SIR FREDERICK, *Thinking*, Allen & Unwin, London, 1958.

BELL, E. T., *Men of Mathematics*, Penguin Books, Harmondsworth, 1953.

BRUNER, J. S., *The Process of Education*, Harvard University Press, Cambridge, Massachusetts, 1965.

BUXTON, L. G., *Do you Panic about Maths?*, Heinemann Educational, London, 1981.

DIENES, Z. P., *An Experimental Study of Mathematics Learning*, Hutchinson, London, 1963.

PIAGET, J., *The Child's Conception of Number*, Routledge & Kegan Paul, London, 1952.

SAWYER, W. W., *Mathematician's Delight*, Penguin Books, Harmondsworth, 1943.

Prelude to Mathematics, Penguin Books, Harmondsworth, 1955.

SKEMP, R. R., *Intelligence, Learning, and Action*, Wiley, New York, 1979.

Index

Abstracting, 21, 223
Abstraction, 21, 90
Accuracy, 67 ff
Activity
  automatic, 83 ff
  creative, 85 ff
  higher mental, 119
  mechanical, 83
  reflective, 77 ff
  reflective, and anxiety, 119 ff
Adaptability, 41, 44
Adding, 149 ff, 179 ff, 189, 196 ff,
  208
Addition
  associative property of, 158 ff
  commutative property of, 158 ff
Algebra, 215 ff
Ambiguity, avoidance of, 69 ff
Analysis, conceptual, 32
Angles, notation for, 79 ff
Anxiety, 118 ff
  adaptations to, 122 ff
  and higher mental activity, 119 ff
Arguments, visually presented, 95 ff
Assimilation, 43 ff, 212 ff
  difficulties of, 48 ff
Associative property, 59, 158 ff, 185
Authority, 109, 112 ff

Basis, 271 ff, 274
Binary
  numbers, so-called, 65 ff
  numerals, 66 ff
  operation, 185
Blockage, mental, 121
Bruner, 87

Cantor, 205
Carrying, 153
Cartesian coordinates, 254 ff
Characteristic property, of a set,
  134 ff, 145
Circle
  equation of, 101
  graph of, 257
Circles, vicious, in learning, 121 ff
Classification, multiple, 75
Classifying, 21
Cognitive strain, 79
Communication, 64 ff
  accuracy of, 67 ff
Commutative property, 158 ff
Concept, 19 ff
Concepts
  communication of, 23 ff, 30, 73 ff
  consciousness of, 78
  as a cultural heritage, 26 ff
  detachability from experience, 26 ff
  higher- and lower-order, 25
  mathematical, 30 ff
  primary, 24
  secondary, 24
Conceptual
  analysis, 32
  hierarchy, 25
  thinking, power of, 29 ff
Context, 70 ff
Coordinates, 100 ff, 254 ff
Correspondence, one-to-one, 138 ff,
  145, 233 ff
Counting, 133 ff, 141 ff, 146
Criterion, for validity of statement,
  19 ff

Dedekind, 205
Derivative, 249
Descartes, 100, 254 ff
Differentiation, 248
Digit, 153
Direction, 259 ff
Discipline, 113
Discourse, universe of, 135
Discussion, value of, 61 ff, 92, 114 ff
Distributive property, 156
Domain, of function, 243

Element, 134, 145
Ellipse, 101, 258
Embodying, 223
Emotional factors, 108
Enlargement, 244
Equality, 165 ff
Equation, 224 ff
    of circle, 101
    of ellipse, 101
    of hyperbola, 102
    of parabola, 102
    solution of, 49 ff, 225
Equations, equivalent, 227
Equivalence, 162 ff
    class, 164, 260 ff
    relation, 36 ff, 163
    relation, properties of, 172
Equivalent
    equations, 227
    ratios, 195 ff
Expansion, 212
Explanation, 76 ff

Figure, geometric, 250 ff
Formula, 84
Formulating, 56
Fractional
    numbers, 178 ff
    number system, 183 ff
Fractions, 174 ff
    equivalent, 177 ff
Frame of reference, 253
Function, 36, 103
    algebraic, 243

    domain of, 243
Functions
    composition of, 246
    operations on, 244

Galton, 88
Generalization, mathematical, 55 ff
Geometry, 250 ff
    Euclidean, 250 ff
Ghiselin, 86
Glennon, 106 ff
Graph, Cartesian, 242, 256 ff
Groups
    attitudes within, 116
    mathematical, 82

Hemispheric specialization, 106
Hierarchy, conceptual, 25
Hyperbola, equation of, 102

Iff, meaning of, 138
Imagery, 88
    visual and verbal compared, 103 ff
Implication, 96, 236
Index notation, 56 ff, 216 ff
Information, recovery of, 84 ff
Insight, 86
Insults, to intelligence, 110 ff
Integers, 189 ff
    alternative notation for, 192 ff
    as a number system, 191 ff
Intelligence
    A, 17
    B, 16 ff
    insults to, 110 ff
    intuitive, 51 ff
    reflective, 51 ff, 60
Interchangeability principle, 164 ff
Interpersonal factors, 108
Intervals, nests of, 205 ff
Isomorphism, 187 ff, 235

Knowledge, recording, 68 ff

Learning
    habit, 15 ff

intelligent, 15 ff
motivations for, 123 ff
rote, 15 ff, 40
schematic, 40 ff
Line, 251
Loci, 254
Logical presentation, 13

Manipulations, routine, 82 ff
Mappings, 231 ff
Matching, 137 ff, 163
Mathematical model, 162 ff, 167 ff
Meaning, 65
Measurement, 168 ff
Memorizing, 48
Mixed numbers, 186 ff, 193, 210 ff
Mnemonic, 84
Model, mathematical, 162 ff, 167 ff
Modular arithmetic, 82
Motivation, 118 ff, 123 ff
  extrinsic, 124
  intrinsic, 124 ff
Multiplication, 154, 180, 188, 197,
  208
  associative property of, 159 ff
  commutative property of,
  159 ff

Naming, 22 ff
  of numbers, 147 ff
Natural numbers, 42, 166
Natural number system, properties of,
  160 ff
Nests, of intervals, 205
Networks, topological, 44 ff
Newton, 27, 136
Noise, 28
Notation
  for angles, 79 ff
  mathematical, 79 ff
Number, 133, 145
  generalization of idea, 58
Numbers
  counting, 42
  fractional, 178 ff
  irrational, 211

mixed, 186 ff, 193
naming of, 147 ff
natural, 42, 166
rational, 194 ff
real, 203 ff
Number systems
  defined, 185
  formal properties of, 59
Numerals, 66 ff, 147 ff
  binary, 66 ff
  Roman, 147 ff
Numeration, 147 ff
  Hindu-Arabic system, 147 ff

Ohm's law, 85
One-to-one correspondence, 138 ff,
  145, 232 ff
Orthogonal projection, 240

Parabola, 101, 258
Piaget, 53, 90, 133, 144
Points, 250 ff
Point sets, 253
Position, 252
Problem-solving, 213 ff
Projection, orthogonal, 240
Properties of number systems, 160 ff,
  183 ff
Property
  associative, 59, 158 ff, 185
  characteristic, of set, 134, 145
  commutative, 59, 159 ff, 185
  distributive, 156, 185
  transitive, 139 ff, 163
Proportion, 196
Psychological presentation, 13
Pythagoras, 203 ff

Ratio, 196 ff
Rational numbers, 199 ff
  in fractional notation, 200 ff
Ratios
  equivalent, 195
  negative, 198
Realms of thought, 171 ff
Real numbers, 202 ff

Recognition, 19 ff, 26
Reconstruction, 41 ff
    difficulty of, 42 ff
Recording, 68
Recovery
    of information, 84 ff
    of understanding, 84 ff
Red, concept of, 137
Reference, frame of, 253
Reflection, 243
Reflective
    activity, 77 ff
    intelligence, 51 ff, 60
Relation
    equivalence, 36, 162
    mathematical, 35 ff
    order, 36
Rigour, 67
Rotation, 244
Rules, 111 ff

Schema, 16, 35 ff
    integrative function of, 37 ff
    selective effect of, 41
    as tool for learning, 38
Schemas
    deliberate changes in, 51 ff
    short- and long-term, 48 ff
Set, 133 ff, 145
    characteristic property of, 134, 145
    truth, 225
Sets
    matching, 137 ff
    ordered, 141
    of points, 253 ff
    standard, 140 ff, 146
    union of, 149 ff
Shear, 244
Similar triangles, 269
Space, 250 ff
    vector, 278 ff
Speech
    audible, 91 ff
    sub-vocal, 91
Strain, cognitive, 79
Stretch, 244

Structure, 78 ff
Sum, 150 ff
Symbols
    functions of, 64 ff
    verbal, 88 ff
    visual, 88 ff
System, mathematical, 234

Tally, 140
Tasks
    complexity of, 118
    of teacher, 63, 108 ff
Teacher
    authoritarian, 121
    as group leader, 116 ff
    responsibility of, 49
    roles of, 113
Teaching, 34
Thinking
    conceptual, 29
    socialized, 92 ff
    verbal, 78
Thought, realms of, 171 ff
Three, concept of, 136 ff
Topological networks, 44 ff
Transformation, 36
    geometric, 243 ff
Transitive property, 139 ff, 163
Translation, 244
Traverse, 44
Triangles
    right-angled, 203 ff
    similar, 269

Understanding, 14, 43 ff
    recovery of, 84 ff
Union, 150 ff
Universe of discourse, 135

Variable, 213 ff
    numerical, 215 ff
    value of, 214
Vectors
    addition of, 262 ff
    base, 267
    free, 261

generalized, 274 ff
geometric, 259 ff
localized, 262
multiple by a number, 264
position, 267
Vector spaces, 278 ff
Vertices, 44
   order of, 45

Visually presented arguments, 95 ff
Volta, 136
Vygotsky, 22

Weighing, 169 ff

Yerkes–Dodson law, 118 ff

FOR THE BEST IN PAPERBACKS, LOOK FOR THE 🐧

In every corner of the world, on every subject under the sun, Penguin represents quality and variety – the very best in publishing today.

For complete information about books available from Penguin – including Pelicans, Puffins, Peregrines and Penguin Classics – and how to order them, write to us at the appropriate address below. Please note that for copyright reasons the selection of books varies from country to country.

In the United Kingdom: For a complete list of books available from Penguin in the U.K., please write to *Dept E.P., Penguin Books Ltd, Harmondsworth, Middlesex, UB7 0DA*

In the United States: For a complete list of books available from Penguin in the U.S., please write to *Dept BA, Penguin, 299 Murray Hill Parkway, East Rutherford, New Jersey 07073*

In Canada: For a complete list of books available from Penguin in Canada, please write to *Penguin Books Canada Ltd, 2801 John Street, Markham, Ontario L3R 1B4*

In Australia: For a complete list of books available from Penguin in Australia, please write to the *Marketing Department, Penguin Books Australia Ltd, P.O. Box 257, Ringwood, Victoria 3134*

In New Zealand: For a complete list of books available from Penguin in New Zealand, please write to the *Marketing Department, Penguin Books (NZ) Ltd, Private Bag, Takapuna, Auckland 9*

In India: For a complete list of books available from Penguin, please write to *Penguin Overseas Ltd, 706 Eros Apartments, 56 Nehru Place, New Delhi, 110019*

In Holland: For a complete list of books available from Penguin in Holland, please write to *Penguin Books Nederland B.V., Postbus 195, NL–1380AD Weesp, Netherlands*

In Germany: For a complete list of books available from Penguin, please write to *Penguin Books Ltd, Friedrichstrasse 10 – 12, D–6000 Frankfurt Main 1, Federal Republic of Germany*

In Spain: For a complete list of books available from Penguin in Spain, please write to *Longman Penguin España, Calle San Nicolas 15, E–28013 Madrid, Spain*

THE PENGUIN ENGLISH DICTIONARY

The Penguin English Dictionary has been created specially for today's needs. It features:

* More entries than any other popularly priced dictionary
* Exceptionally clear and precise definitions
* For the first time in an equivalent dictionary, the internationally recognised IPA pronunciation system
* Emphasis on contemporary usage
* Extended coverage of both the spoken and the written word
* Scientific tables
* Technical words
* Informal and colloquial expressions
* Vocabulary most widely used *wherever* English is spoken
* Most commonly used abbreviations

It is twenty years since the publication of the last English dictionary by Penguin and the compilation of this entirely new *Penguin English Dictionary* is the result of a special collaboration between Longman, one of the world's leading dictionary publishers, and Penguin Books. The material is based entirely on the database of the acclaimed *Longman Dictionary of the English Language.*

1008 pages   051.139 3   £2.50 ☐

FOR THE BEST IN PAPERBACKS, LOOK FOR THE 🐧

PENGUIN DICTIONARIES

Archaeology

Architecture

Art and Artists

Biology

Botany

Building

Chemistry

Civil Engineering

Commerce

Computers

Decorative Arts

Design and Designers

Economics

English and European
  History

English Idioms

Geography

Geology

Historical Slang

Literary Terms

Mathematics

Microprocessors

Modern History 1789–1945

Modern Quotations

Physical Geography

Physics

Political Quotations

Politics

Proverbs

Psychology

Quotations

Religions

Saints

Science

Sociology

Surnames

Telecommunications

The Theatre

Troublesome Words

Twentieth Century History

Dictionaries of all these – and more – in Penguin

FOR THE BEST IN PAPERBACKS, LOOK FOR THE 🐧

A CHOICE OF PENGUINS AND PELICANS

Lateral Thinking for Management   Edward de Bono

Creativity and lateral thinking can work together for managers in developing new products or ideas; Edward de Bono shows how.

Understanding Organizations   Charles B. Handy

Of practical as well as theoretical interest, this book shows how general concepts can help solve specific organizational problems.

The Art of Japanese Management   Richard Tanner Pascale and Anthony G. Athos   With an Introduction by Sir Peter Parker

Japanese industrial success owes much to Japanese management techniques, which we in the West neglect at our peril. The lessons are set out in this important book.

My Years with General Motors   Alfred P. Sloan   With an Introduction by John Egan

A business classic by the man who took General Motors to the top – and kept them there for decades.

Introducing Management   Ken Elliott and Peter Lawrence (eds.)

An important and comprehensive collection of texts on modern management which draw some provocative conclusions.

English Culture and the Decline of the Industrial Spirit   Martin J. Wiener

A major analysis of why the 'world's first industrial nation has never been comfortable with industrialism'. 'Very persuasive' – Anthony Sampson in the *Observer*

FOR THE BEST IN PAPERBACKS, LOOK FOR THE 🐧

A CHOICE OF PENGUINS AND PELICANS

Dinosaur and Co   Tom Lloyd

A lively and optimistic survey of a new breed of businessmen who are breaking away from huge companies to form dynamic enterprises in microelectronics, biotechnology and other developing areas.

The Money Machine: How the City Works   Philip Coggan

How are the big deals made? Which are the institutions that *really* matter? What causes the pound to rise or interest rates to fall? This book provides clear and concise answers to these and many other money-related questions.

Parkinson's Law   C. Northcote Parkinson

'Work expands so as to fill the time available for its completion': that law underlies this 'extraordinarily funny and witty book' (Stephen Potter in the *Sunday Times*) which also makes some painfully serious points for those in business or the Civil Service.

Debt and Danger   Harold Lever and Christopher Huhne

The international debt crisis was brought about by Western bankers in search of quick profit and is now one of our most pressing problems. This book looks at the background and shows what we must do to avoid disaster.

Lloyd's Bank Tax Guide 1986/7

Cut through the complexities! Work the system in *your* favour! Don't pay a penny more than you have to! Written for anyone who has to deal with personal tax, this up-to-date and concise new handbook includes all the important changes in this year's budget.

The Spirit of Enterprise   George Gilder

A lucidly written and excitingly argued defence of capitalism and the role of the entrepreneur within it.

FOR THE BEST IN PAPERBACKS, LOOK FOR THE 🐧

A CHOICE OF PENGUINS AND PELICANS

Asimov's New Guide to Science Isaac Asimov

A fully updated edition of a classic work – far and away the best one-volume survey of all the physical and biological sciences.

Relativity for the Layman James A. Coleman

Of this book Albert Einstein said: 'Gives a really clear idea of the problem, especially the development of our knowledge concerning the propagation of light and the difficulties which arose from the apparently inevitable introduction of the ether.

The Double Helix James D. Watson

Watson's vivid and outspoken account of how he and Crick discovered the structure of DNA (and won themselves a Nobel Prize) – one of the greatest scientific achievements of the century.

Ever Since Darwin Stephen Jay Gould

'Stephen Gould's writing is elegant, erudite, witty, coherent and forceful' – Richard Dawkins, *Nature*

Mathematical Magic Show Martin Gardner

A further mind-bending collection of puzzles, games and diversions by the undisputed master of recreational mathematics.

Silent Spring Rachel Carson

The brilliant book which provided the impetus for the ecological movement – and has retained its supreme power to this day.

FOR THE BEST IN PAPERBACKS, LOOK FOR THE 🐧

A CHOICE OF PENGUINS AND PELICANS

Setting Genes to Work   Stephanie Yanchinski

Combining informativeness and accuracy with readability, Stephanie Yanchinski explores the hopes, fears and, more importantly, the realities of biotechnology – the science of using micro-organisms to manufacture chemicals, drugs, fuel and food.

Brighter than a Thousand Suns   Robert Jungk

'By far the most interesting historical work on the atomic bomb I know of' – C. P. Snow

Turing's Man   J. David Bolter

We live today in a computer age, which has meant some startling changes in the ways we understand freedom, creativity and language. This major book looks at the implications.

Einstein's Universe   Nigel Calder

'A valuable contribution to the de-mystification of relativity' – *Nature*

The Creative Computer   Donald R. Michie and Rory Johnston

Computers *can* create the new knowledge we need to solve some of our most pressing human problems; this path-breaking book shows how.

Only One Earth   Barbara Ward and Rene Dubos

An extraordinary document which explains with eloquence and passion how we should go about 'the care and maintenance of a small planet'.